The Journal of Practical Ecology and Conservation
Volume 7 No. 2 2009

Published by:

Wildtrack Publishing
Venture House
103 Arundel Street,
Sheffield, S1 2NT
UK

Edited by Ian D. Rotherham

ISSN 1354-0270

ISBN 978-1-904098-44-7

Front cover picture: Woodhouse Mill area, Sheffield UK. © Christine Handley

Journal of Practical Ecology and Conservation Volume 7(2) 2009

Contents

Assessing impacts and sensitivities of proposed flood defence schemes a landscape sensitivity and capacity approach

Peter Glaves
University of Northumbria

Editorial Paper

Abstract

Flood management schemes have economic, social and environmental impacts which operate over large areas. Methods are needed to holistically assess such impacts. Landscape character assessments and landscape impact assessments provide standardised methodologies for characterising and assessing environments (both socially and ecologically). Such methods have been successfully used to assess the risk of new developments and the capacity of areas to accept developments. Such methods can be successfully adapted to assess the significance of proposed flood management schemes on a landscape scale.

Introduction

The traditional response to increasing flood risk in coastal and riverine environments has been hard defences. However, as the threat of human-induced climate change and sea-level rise become more pronounced, a wider range of adaptation options have been considered (Shih and Nicholls, 2007), including flood storage, realignment, *etc*.

Increasingly the emphasis in flood management is on large scale strategic and sustainable approaches, for example the Environment Agency River Severn Project and the Thames Estuary 2100 project, the Rhine and Meuse river basins, *etc*. Such approaches consider a range of flood management options on a regional or landscape scale. The benefits and risks of such schemes have been well documented. The impacts of new flood management have been assessed in terms of biodiversity, habitats, hydrology and land uses (e.g. Burd, 1995; Myatt *et al.*, 2003, CIRIA, 2004).

The UK Government's flood policy is to reduce the risks to people and the natural environment by encouraging *"the provision of technically, environmentally and economically sound and sustainable defence measures"* (RPA, 2006). Many flood related studies, however, have focused on impacts in small geographical areas, and/or on specific types of impacts. There is a need for approaches that consider the full range of impacts, and which are socially, economically and environmentally sustainable and sound.

Landscape character assessment and sensitivity analysis approaches have been used to characterise environments on a landscape scale and to assess the potential sensitivities in relation to regional planning (e.g. Essex – Chris Blandford Associates, 2003; High Peak – Countryscape *et al.*, 2006) and to large scale schemes, including wind farms (e.g. Land Use Consultants, 2005; Scottish Natural Heritage, 2005 & 2006). These are established techniques which consider the full range of issues operating within a region and are therefore potentially applicable to assessing impacts of flood management schemes.

This paper presents a review of the potential use of landscape sensitivity and capacity approaches as an approach for an integrated assessment of the impacts of flood defence schemes. It begins by explaining what landscape is and the standard approach to its assessment and characterisation in the UK (Countryside Agency, 2002). It then goes on to explain how the capacity of an area to proposed flood management options (walls, realignment, flood storage, barriers and barrages) can be determined and the sensitivities identified.

Landscape Defined

Landscape is a term that has been much defined. Traditional definitions come from either an ecological and physical standpoint or from an artistic and aesthetic outlook. As stated in the Landscape Character Assessment Guidance (Swanwick for the Countryside Agency and Scottish Natural Heritage, 2002):

Landscape is about the relationship between people and place. It provides the setting for our day-to-day lives. The term does not mean just special or designated landscapes and it does not only apply to the countryside. Landscape can mean a small patch of urban wasteland as much as a mountain range, and an urban park as much as an expanse of lowland plain. It results from the way that different components of our environment - both natural (the influences of geology, soils, climate, flora and fauna) and cultural (the historical and current impact of land use, settlement, enclosure and other human interventions) - interact together and are perceived by us.

A landscape therefore contains all the elements (physical, biological, social, economic and perception) which make up the characteristics of a geographical area and therefore all the elements which will be affected by developments such as flood defences. An assessment of flood management impacts which uses a landscape-based approach, therefore, will be more likely to be holistic and strategic in its approach.

Landscape is therefore a term for all the components that make up countryside and settlements; it can be used as a synonym for the environment. Landscape assessment of flood management is therefore an assessment of the environmental impact of the scheme at a landscape scale.

The important values of landscapes, and therefore the reasons for assessing impacts on landscapes, are set out in the European Landscape Convention (2000), specifically the need to consider landscapes in development because:

- of its interest to the public - culturally, ecologically, environmentally and socially;

- it is a resource favourable to economic activity and whose protection, management and planning can contribute to job creation;

- of its importance to local cultures;

- it is a basic component of the European natural and cultural heritage, contributing to human well-being and consolidation of the European identity;

- it is an important part of the quality of life for people everywhere: in urban areas and in the countryside, in degraded areas as well as in areas of high quality, in areas recognised as being of outstanding beauty as well as everyday areas;

- it is a key element of individual and social well-being; and

- its protection, management and planning entail rights and responsibilities for everyone.

(extract from the Preamble to the European Landscape Convention, 2000).

Clearly, therefore, landscapes have a relevance to flood management schemes. The following sections set out the standard methods and approaches by which landscapes are characterised and assessed.

Landscape Assessment and Characterisation

Landscape assessment is a tool for identifying and describing the character of our landscape and for recognising the key management and conservation issues (CRC, 1995). The approach first emerged in the mid-1980s and since 1993 has included a specific emphasis on landscape character as a central concept (Swanwick, 2002) and is known as Landscape Character Assessment (LCA). For details of the history and development of landscape characterisation assessment approaches in the UK see Jensen (2004).

LCA is an approach based on perception and understanding the intrinsic character of the English landscape and its distinctive features (Swanwick, 2002). It is a method of describing the landscape as an environmental resource and analysing how it can be conserved and managed (Landscape Design Associates, 2001) especially focusing on features which create local distinctiveness.

Today it is generally agreed that any strategy for environmental change should be based upon a clear understanding of the characteristics and value of the resource which it seeks to conserve and improve (CRC, 1995). For example, sustainable flood and coastal defence projects need to consider the following elements:

- Preserving and enhancing the environment;

- Using resources efficiently;

- Ensuring design, operation, and maintenance processes are efficient and flexible to long-term needs (MAFF, 2000).

LCAs can provide the basis for an assessment of the sustainability of a flood management scheme and provide information on the current area characteristics and values and the impacts of the proposals.

A landscape's character is physically made up of the distinct and recognisable pattern of elements/features that occur within that particular type of landscape. Landscape character is also determined by how these elements are perceived by people, creating the particular sense of place of different areas of the landscape. LCAs, therefore, consider the character of an area now and the impacts of development both environmentally and socially.

A landscape's character is made up of both features and factors. Factors are circumstances or influences that contribute to the impression of a landscape (e.g. scale, enclosure, elevation). Landscape features are prominent eye-catching elements, e.g. wooded hill top or church spire. Together these elements create the character of that landscape.

LCA does not just list landscape character and features, etc., it also considers landscape quality or condition, i.e. it makes judgments about the physical state of the landscape and about its intactness, from visual, functional and ecological perspectives. It also reflects the state of repair of individual features and elements which make up the character in any one place.

The overall combination of the above elements that contribute to landscape context, character and value are referred to as the landscape resource. It is this resource which will be affected by developments such as flood management. The resource therefore has to be characterised and evaluated before any impacts of development can be assessed. Landscape characterisation, therefore, forms the first stage in assessing the impacts of flood management. The elements considered in a LCA are shown in Figure 1.

Figure 1. Elements of Landscape Characterisation

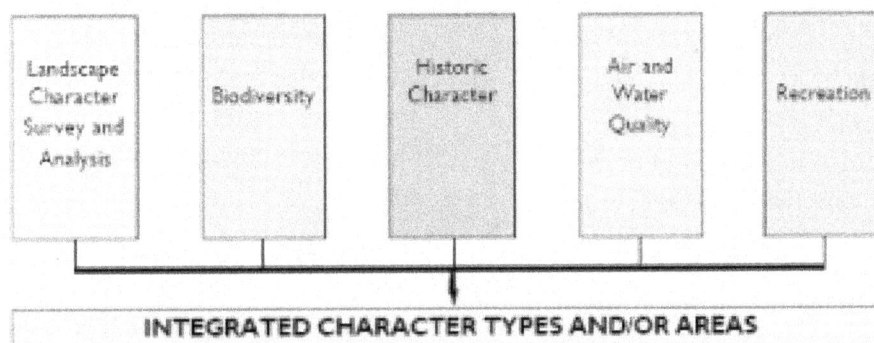

Landscape Character Survey and Analysis | Biodiversity | Historic Character | Air and Water Quality | Recreation

INTEGRATED CHARACTER TYPES AND/OR AREAS

Countryside Agency (2002)

An analysis of the above landscape elements enables a LCA to divide up a geographical area into landscape character types/areas, i.e. areas of land having broadly similar patterns of geology 'landform' soils, vegetation, land use, settlement and field patterns. Flood management within a single landscape character area will be impacting on a common set of environmental and social characteristics. Landscape character areas can be created at a series of scales from local to regional depending on the scale of the flood management scheme and the detail of the scheme. Landscape character areas offer a suitable approach for dividing up a study area into different zones with different characteristics, values and impact types.

Landscape characterisation is used as the basis of landscape evaluation, the landscape value being the relative value or importance attached to a landscape. Criteria used in assessing value include condition, significance and robustness (see Countryside Agency, 2004 - Topic paper 6 in particular s.4.2. for details). Such landscape values are recognized in designations. Values may include scenic beauty, tranquillity or wildness, cultural associations or other conservation issues.

A LCA provides an analysis of the current landscape resource; this forms the basis for assessing impacts of flood management on this resource, this process being known as landscape impact assessment.

Landscape Impact Assessment

Landscape Impact Assessment considers the effects of development on the landscape, specifically changes in the elements, characteristics, character and qualities of the landscape as a result of development. These effects can be positive or negative.

Condition, Sensitivity and Capacity are key elements in determining landscape impacts, each of these words are defined in current guidance (Landscape Institute, 2002 and Countryside Agency, 2004).

Landscape condition is strongly influenced by the impact of external factors. The assessment of condition evaluates the

pattern of the landscape and the presence of incongruous features on the unity of the landscape. It also evaluates how well the landscape functions as a habitat for flora and fauna and the condition of cultural or 'man-made' elements such as enclosures, built elements and roads.

Condition is defined by an analysis of visual unity and functional integrity and is classified as very poor, poor, moderate, good and very good (*ibid*).

Landscape sensitivity is a measure of the ability of a landscape to accept change without causing irreparable damage to the essential fabric and distinctiveness of that landscape. The term 'change' refers to both beneficial changes, such as a new woodland, as well as change that may be brought about by new land uses. Sensitivity is defined by an analysis of *Sense of Place* and *Visibility* and ranges from very low through low, moderate, high to very high (*ibid*).

Landscape capacity measures the extent to which a particular landscape is able to accommodate change relative to a specific proposal, such as flood management.

Judging landscape character sensitivity requires professional judgement about the degree to which the landscape in question is robust, in that it is able to accommodate change without adverse impacts on character. This involves making decisions about whether or not significant characteristic elements of the landscape will be liable to loss, and whether important aesthetic aspects of character will be liable to change (Countryside Agency, 2006 - Topic Paper 6 - techniques and criteria for judging capacity and sensitivity para. 4.2).

Further details on the assessment of landscape sensitivity and capacity are given in Swanwick (2004). The above assessment links to the requirements of current guidance, specifically:

Visual and Landscape Impact Assessments (VIA and LIA) have become a statutory requirement of Environment Assessment (EA) within the UK (see Colville, 1996). The starting point for a LIA is a full landscape

Figure 2. Visual and Landscape Impacts (adapted from Landscape Institute and Institute of Environmental Assessment 2002)

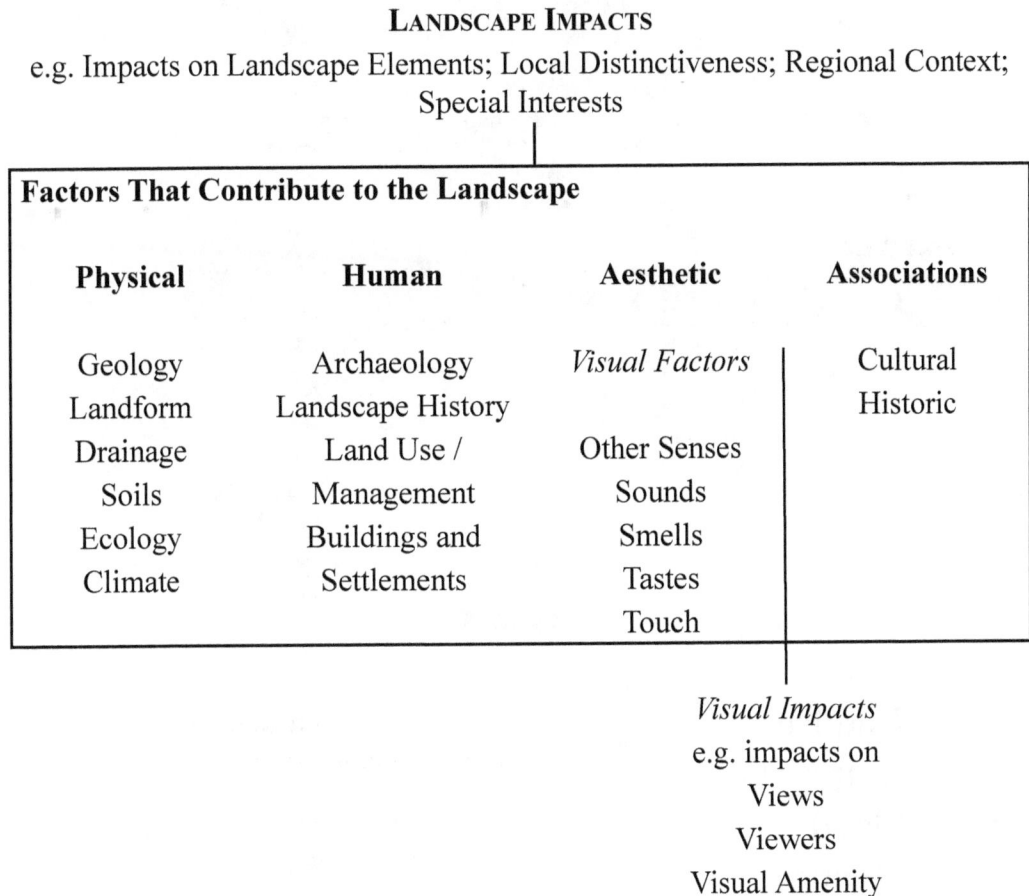

LANDSCAPE IMPACTS

e.g. Impacts on Landscape Elements; Local Distinctiveness; Regional Context; Special Interests

Factors That Contribute to the Landscape			
Physical	**Human**	**Aesthetic**	**Associations**
Geology	Archaeology	*Visual Factors*	Cultural
Landform	Landscape History		Historic
Drainage	Land Use /	Other Senses	
Soils	Management	Sounds	
Ecology	Buildings and	Smells	
Climate	Settlements	Tastes	
		Touch	

Visual Impacts
e.g. impacts on
Views
Viewers
Visual Amenity

characterisation or landscape character assessment (LCA) as detailed above. A LIA firstly considers the current landscape characteristics, it considers its value and sensitivities and then goes on to assess the landscape and visual impacts of development on this character, for example the developments and change in land use associated with the flood management options.

A landscape impact assessment covers both landscape impacts, i.e. changes in the elements and character of the landscape, and visual impacts, for example changes in views of the landscape and effects of changes on people. Landscape and visual impacts are closely interrelated (as shown in Figure 2).

Landscape Impacts are changes in the fabric, character and quality of the landscape as a result of a development, both directly and indirectly, as well as impacts on specific features and overall character. *Visual Impacts*

relate solely to changes in available views of the landscape and the effects of those changes on people.

Assessing landscape impact involves systematically identifying all the potential landscape and visual impacts associated with the development, predicting their magnitude and significance. The significance of landscape or visual impacts is a function of the sensitivity of the affected landscape and the magnitude of change that they will experience (see Figure 3).

Moderate and substantial impacts are regarded as being significant or very significant (respectively). Mitigation has to be considered for any significant landscape or visual impact of development.

Figure 3. Significance of Landscape Impacts (adapted from Landscape Institute and Institute of Environmental Assessment, 2002)

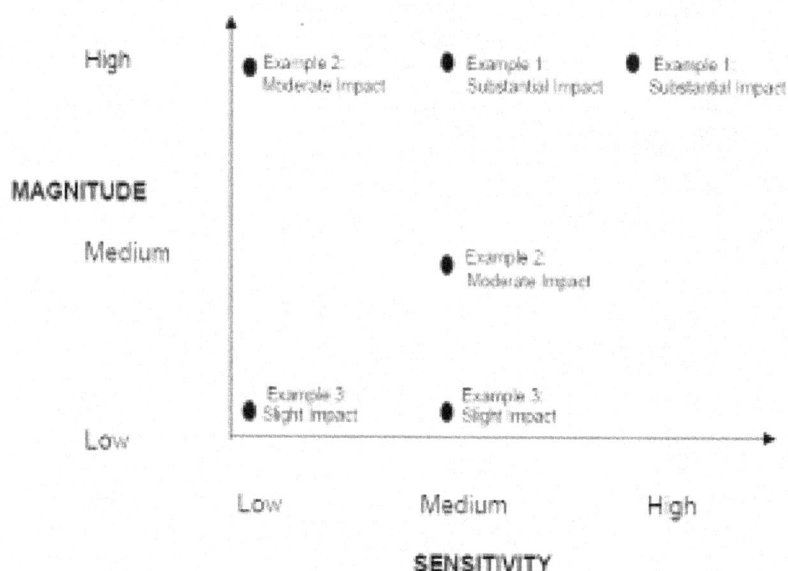

Application of Landscape Sensitivity and Capacity Assessment to Flood Management Schemes

There are clear landscape and visual impacts related to the different flood management options (walls, barriers, barrages, realignment, *etc*). These include visual impacts, changes in land cover and landscape character and quality. Landscape impact issues include major wall-raising and enclosure of riverside landscapes, the visual intrusion of river structures, and the need for careful design. The impact of separating populations from the estuary by raised sea defences can affect how people value and use the estuary, and affect their perception of flood risk and their safety. In more rural areas the landscape and landscape ecological impacts result from water storage, realignment and wall-raising.

The Hampshire County Council (HCC, 2005) approach to landscape sensitivity assessment provides an appropriate approach for considering the landscape sensitivity/ capacities relating to proposed flood management schemes, and has been adapted to assess the potential impacts of the Environment Agency Thames Estuary 2100 project.

The Hampshire approach (*ibid*) considers impacts relating to five themes, i.e.:

- **The physical landscape**, covering soils, landform and land cover;

- **The experiential landscape**, covering ruralness, tranquillity and access;

- **Biodiversity**, with reference to both common and rare habitats and species and their designations;

- **Historic environment**, addressing archaeology, built environment and historic landscape; all of which contribute to landscape character sensitivity; and

- **Visibility**, covering physical prominence, enclosure or openness, zones of visual influence and types of view.

For each theme a series of attributes relating to landscape features and characteristics have been defined, each being assessed against three indicators to establish the extent to which they are sensitive to the proposed flood management. The individual findings aggregated up to establish a level of sensitivity and to assess the capacity of the landscape to cope with the proposed flood defences.

Figure 4. Assessment of Landscape Significance, Robustness and Condition (HCC, 2005)

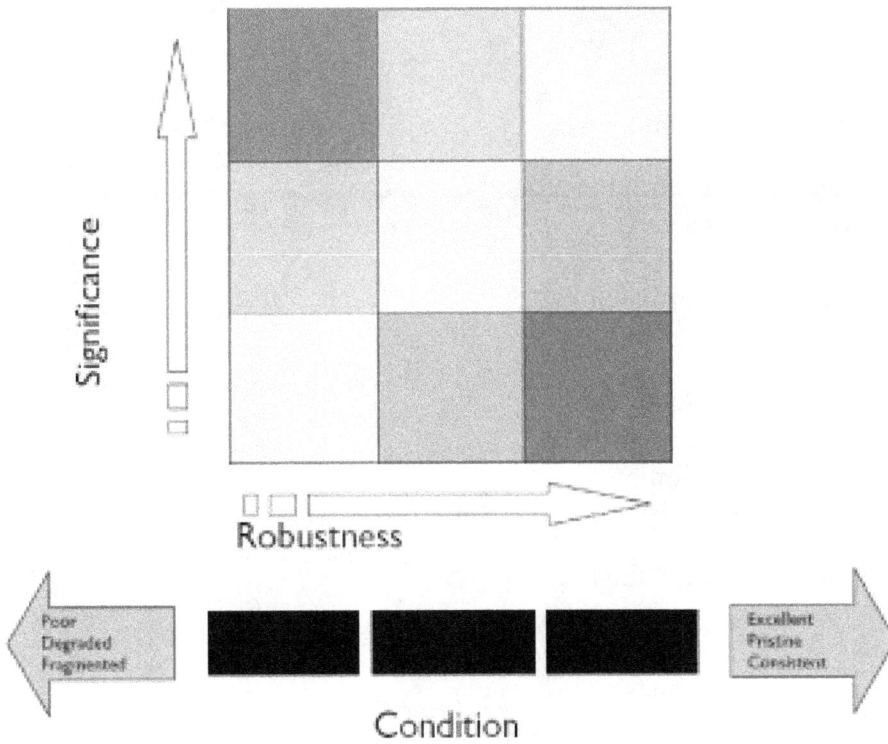

Figure 5. Diagram illustrating how overall sensitivity/capacity can be calculated by overlaying individual sensitivities for landscape elements (HCC, 2005)

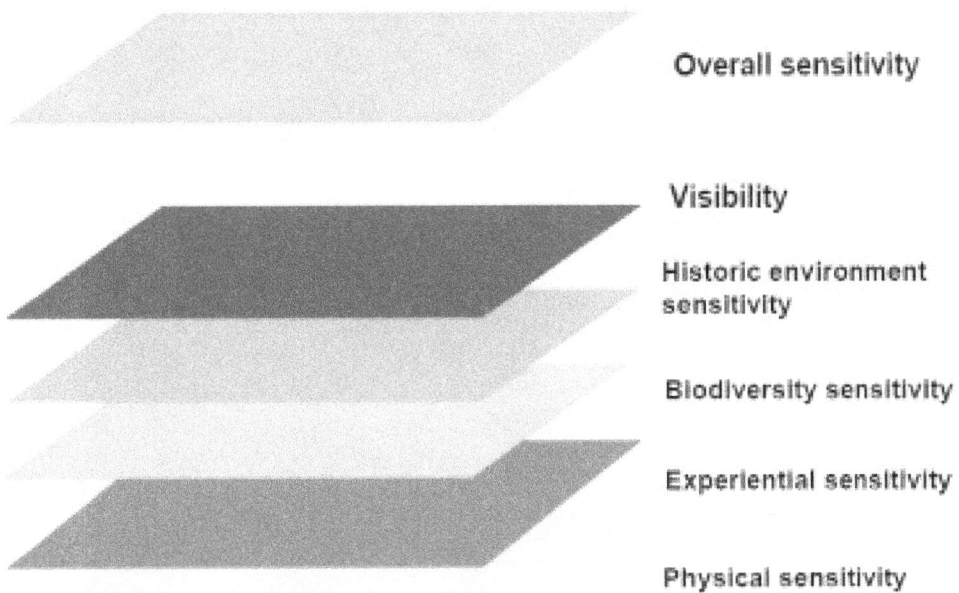

In this method, three indicators are used to define sensitivity/capacity:

- **Significance** - an indication of rarity, and representativeness of that attribute within the landscape;

- **Robustness** - the attribute's vulnerability and fragility; and

- **Condition** - how well the attribute has been preserved, conserved and managed (see Countryside Agency Topic Paper 6 (2004) in particular s.4.2 for more details).

Following the Hampshire method, the significance and robustness of attributes are recorded on a matrix, and condition on a continuum, as shown in Figure 4.

The results for each attribute are then compiled to determine overall sensitivities for each theme. The themes are in turn combined (Figure 5) to produce an overall landscape sensitivity and capacity rating.

The approach enables the assessment of the sensitivity of the landscape as a whole and of its component elements, in terms of its overall character, its quality and condition, the aesthetic aspects of its character and also the sensitivity of individual elements contributing to the landscape. The approach therefore meets the current guidance as set out in Countryside Agency (2004, paras 4.1 and 4.2).

The Hampshire approach has been adapted to assess the capacity of the landscape to cope with the proposed flood management and to identify any attributes that will be significantly affected. It does this by assessing the:

- type of flood management;

- landscape character; and

- landscape sensitivities.

The landscape capacity has been summarised in terms of level of risk:

- High risk – low landscape capacity to cope with the proposed flood management option;

- Medium risk – moderate landscape capacity to cope with the proposed flood management option;

- Low risk – high landscape capacity to cope with the proposed flood management option.

The landscape risk approach used is similar to that used in other landscape assessments, e.g Thurrock (undated), see Figure 6.

Figure 6. Thurrock Landscape Risk Criteria

Landscape Sensitivity Level	Sensitivity Criteria	Ability of the Landscape to Absorb Impacts of Development and Other Change
High	The landscape is very sensitive to this type/ scale of development/ change due to the potential for very adverse impacts on: • Distinctive physical and cultural components or key characteristics. • Strength of character / condition of the landscape. • AONB Landscape. • Landscape of high intervisibility/ visual exposure. • Tranquil area. with very limited opportunities for mitigation.	Unlikely to be capable of being absorbed. Presumption against development unless overriding need.
Moderate	The landscape is sensitive to this type/ scale of development/ change due to the potential for some adverse impacts on: • Distinctive physical and cultural components or key characteristics. • Strength of character / condition of the landscape. • Landscape of moderate intervisibility/ visual exposure. • Area of fragmented tranquillity. but there may be more opportunities to overcome these through appropriate siting, design and other mitigation measures.	May be capable of being absorbed. Developments to be considered on their individual merits.
Low	The landscape is less sensitive to this type and scale of development/ change due to the potential for only slight or no damaging impacts on: • Distinctive physical and cultural components or key characteristics. • Strength of character / condition of the landscape. • Landscape of low intervisibility/ visual exposure. • Area with an absence of tranquility. and there are likely to be considerable opportunities for mitigation and/or landscape enhancement.	Likely to be capable in principle of being absorbed.

References

Burd, F. (1995) *Managed retreat: a practical guide.* English Nature, Peterborough.

Chris Blandford Associates (2003) *Essex Landscape Character Assessment.* Essex County Council, Chelmsford.

CIRIA (2004) *Coastal and estuarine managed realignment – design issues.* CIRIA, London.

Colville, K. (1996) *Landscape in Sustainability Appraisal and Strategic Environmental Assessment SA/SEA - Issues and Opportunities.* Landscape Character Network Workshop Landscape Character and Strategic Enviromental Assessment, 13 Sept 2006.

Countryscape and The Planning Cooperative (2006) *From Special Landscape Areas to Landscape Character - Project Methodology Final Report.* High Peak Borough Council, Buxton.

Countryside Agency and Scottish Natural Heritage (2002) *Landscape Character Assessment: Guidance for England and*

Scotland. Countryside Agency CAX 84 available at www.countryside.gov.uk/cci/guidance and www.snh.org.uk/strategy/LCA.

Countryside Agency (2004) *Landscape Character Assessment Guidance for England and Scotland, Topic Paper 6: Techniques and Criteria for Judging Capacity and Sensitivity.* Countryside Agency, Cheltenham.

CRC (Cobham Resource Consultants) (1995) *The Kent Thames Gateway landscape.* CRC, Oxon.

European Landscape Convention (2000). Text available at: http://www.coe.int/t/e/Cultural_Co-operation/Environment/Landscape/.

Hampshire County Council (2005) *Strategic Landscape Sensitivity.* A paper on the working methodology for the Peer Group Workshop, 20th July 2005.

Jensen, K. (2004) *Changing conceptualization of landscape in English landscape assessment methods.* Proceedings of the Frontis workshop from landscape research to landscape planning: aspects of integration, education and application, Wageningen, The Netherlands 1-6 June 2004.

Landscape Design Associates (2001) *Basingstoke and Deane Landscape Assessment,* accessed on line at: www.basingstoke.gov.uk/NR/rdonlyres/2A7533FA-C7B1-41CA-834D-9550D01B003A/0/1Contents_Foreward_Background.pdf.

Landscape Institute and Institute of Environmental Management and Assessment (2002) *Guidelines for Landscape and Visual Impact Assessment,* Second Edition, Spon.

Land Use Consultants (2005) *North Wessex Downs AONB: Landscape Sensitivity to Wind Turbines.* Report for The North Wessex Downs AONB Council of Partners by Land Use Consultants.

MAFF (2000) *Flood and Coastal Defence Project Appraisal Guidance: Environmental Appraisal.* MAFF, London.

Myatt, L.B., Scrimshaw, M.D. and Lester, J.N. (2003) Public perceptions and attitudes towards a current managed realignment scheme: Brancaster West Marsh, north Norfolk, UK. *Journal of Coastal Research,* **19(2),** 278-286.

RPA (2006) *Flood Defence Standards for Designated Sites,* English Nature Research Reports, No. 629, Peterborough.

Scottish Natural Heritage (2005) *Cumulative Effect of Windfarms,* SNH, Edinburgh.

Scottish Natural Heritage (2006) *Visual Representation of Windfarms.* Good Practice Guidance, SNH, Edinburgh.

Shih, C.W. and Nicholls, R.J. (2007) Urban Managed Realignment: Application to the Thames Estuary, London, *Journal of Coastal Research,* **23(6)** 1525-1534.

Swanwick, C. (2004) *Techniques and Criteria for Judging Capacity and Sensitivity,* Landscape Capacity and Sensitivity CCN Workshop, January 2004.

Thurrock District Council (Undated) *Landscape Character Assessment.* Thurrock.

Van Grieten, M, Dower, B and Wigley, K. (2006) *Cumulative landscape and visual impact assessment,* Land Use Consultants, London.

Journal of Practical Ecology and Conservation Volume 7(2) 2009

Sustainable Urban Drainage and Infrastructure

Adrian J. Saul
Pennine Water Group, University of Sheffield

Main Paper

Introduction

The Government's recently published 'Water Strategy for England – Future Water' (2008) - has highlighted the Government's long-term vision for water and the framework for water management in England. Sustainable urban drainage is a key topic in this strategy but significant legislative, infrastructure and behavioural change is likely to be required to promote and apply the strategy. Further, allied to the need to forecast, prevent and better manage urban floods, the strategy, together with the Pitt Review (2007) provide Government and the UK Water Industry with an outstanding opportunity to make a paradigm shift in the way in which water infrastructure is managed, as it may be argued that the current systems in England and Wales are fragmented, inconsistent and sometimes ineffective.

This paper sets the scene for a challenging future where there is a need to move our existing urban drainage systems and infrastructure to better protect the environment and to become more sustainable – technical, environmental, social and economic. Delivery of such sustainable systems is a function of changes in several key drivers, for example, climate change, population growth, carbon and water footprints, changing customer behaviour and perceptions and the need to meet new legislation. Within the urban framework, the balance between the integration of environmental, social and economic issues is delicate, is not fully understood and, at the present time, has no champion.

It is clear that the delivery of more sustainable infrastructure is a wide ranging issue. **In this paper, the urban drainage system and particularly the combined sewer system has been made the focus, with a specific emphasis on urban flooding and the potential infrastructure changes for the future management of urban surface water.** The paper **does not address aspects of the infrastructure used for temporary flood protection, flood resilience of properties or flood repair techniques.**

Types of urban flooding

There are many types of urban flooding and each type results in a different type of surface water on the catchment surface. The sustainable solution and its interaction with infrastructure is also different.

Pluvial Flooding

Rainfall in the urban area may cause flooding due to the fact that there are inadequate hydraulic access pathways to the underground sewer system or due to the fact that the pipes in the sewer system have a hydraulic capacity that is less than the flows that are generated by the rainfall run-off process. In the latter case the sewer pipes are hydraulically inadequate and this results in a back-up of flow and in a 'surcharge' of the system. Such surcharge may result in the internal flooding of basements or external flooding of the catchment surface with the consequent flooding of properties.

Flooding due to Asset Performance, Deterioration or Failure

Sewer flooding is also caused by the performance of assets and asset failure. In many cases such performance is governed by the condition and status of the assets and of the way in which they deteriorate. The primary processes that cause such flooding include:

- Sewer blockages and collapses.

- Presence of sewer sediments.

- Mechanical and equipment failure.

Flooding of the Intra Urban Area from Surrounding Catchments

Such flooding occurs due to a rainfall event on the rural or peri-urban area that surrounds or is adjacent to the intra-urban area. If the flow paths from these surrounding areas lead directly to the intra urban area it is feasible for the surface runoff to flood the intra urban area, in a similar way to floods caused by pluvial flooding.

Flooding due to Fluvial Inundation within Inland Systems

The performance of the sewer system in the intra-urban area may be influenced by the performance of the fluvial drainage system in two ways:

- Hindered performance due to a back-up of flow in the sewer system caused by enhanced fluvial flows that inundate the discharge outlets of the sewer system.

- Inundation of the intra-urban catchment surface due to the failure, overtopping or by-passing of the flood defences of the fluvial system. This results in an inundation of the sewer system that becomes full and subsequently inoperable due to extremely slack hydraulic gradients often with ponding on the low-lying areas of the catchment surface.

Flooding Inundation in Coastal Regions

Similar impacts to those observed in inland systems may be observed in the intra-urban areas adjacent to coastal and estuary environments where, similarly, the performance of the sewer system may be hindered by the height of surges or the overtopping or failure of coastal defences.

Co-incident Flooding

Co-incident flooding occurs when two or more of the above flooding mechanisms occur simultaneously.

The Regulatory Challenge

Defra have a leading role in promoting sustainable drainage and will soon report on the outputs of the Making Space for Water initiative, (Defra, 2004), where the challenges associated with an integrated approach to urban drainage have been assessed. What is clear is that, in respect of surface water management, there are many key stakeholders, including the Defra, EA, Local and Planning Authorities, the Highways Agency, the Water Service providers, Internal Drainage Boards, British Waterways and land owners. The Pitt Review (2007) identified these together with a potential framework as to how these could work together, as shown in Figure 1, with the suggestion that the Environment Agency could be given a strategic role associated with the management of surface water flooding. The challenge of linking bringing together those responsible, together with other Private Companies, Environmental Groups and the General Public should not be underestimated and the debate as to how this may be best managed is ongoing.

In addition a further opportunity to improve the sustainability of our urban drainage infrastructure is the introduction of the *Water Framework Directive* (WFD). Specific aims of the WFD of relevance to this paper include:

- The promotion of sustainable water use;

- The progressive reduction of discharges and emissions;

- The prevention of further deterioration and enhancement of aquatic ecosystems;

- Mitigation of the effect of floods and droughts.

- Ensuring progressive reduction and prevention of groundwater pollution and prevention.

There are many avenues of current research that are ongoing with a view to meeting the needs of the *Water Framework Directive* and some of these are discussed later.

Figure 1. Key Stakeholders with responsibilities in Flood Risk Management.

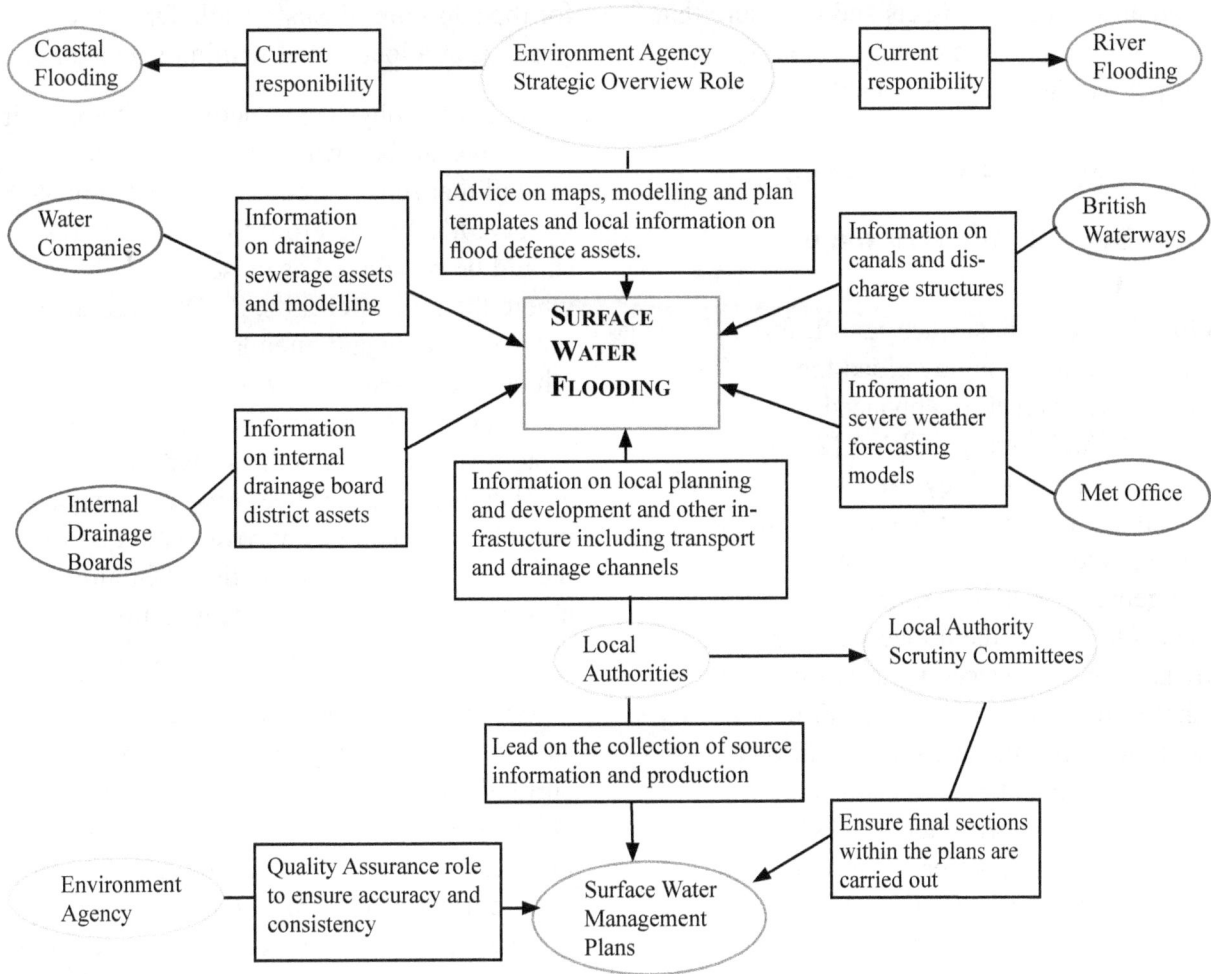

(Crown Copyright 2007, Figure 14, the Pitt Review)

In addition, the introduction of the **Floods Directive** (November 2007), requires that preliminary flood risk assessments are completed to identify areas at risk, with the subsequent development of flood hazard and flood risk maps for areas at risk and the publication of flood risk management plans by December 2015.

Clearly, therefore, there are elements of commonality between these two directives but I am of the opinion that the opportunities to develop sustainable systems offered by the WFD is likely to see significant changes to the surface infrastructure to better manage surface water, and that such changes will have a significant impact on the existing flood risk maps to be prepared under the Floods Directive. For example, as will be shown later, the latest research is promoting significant changes in surface runoff volumes, changes to urban form and the development of safe and reliable urban flood pathways. The introduction of such sustainable infrastructure

is likely to see significant changes to the location of the existing high risk zones in urban areas, and conversely, highlighting a need for close liaison between the two.

In addition, climate change is likely to see increases in rainfall runoff from both pervious and impervious surfaces for some storm events and this, together with increased runoff due urbanisation and urban creep, is likely to see increased runoff into the existing drainage systems. New planning guidance PPS25 sets out Government policy on development and flood risk and the Pitt Review highlighted that Government should look at the need, with legislation in 2008, to change householders' permitted development rights to the 'urban creep' of impermeable surfaces without planning permission.

Hence the Pitt Review highlighted the need for a review of the regulatory framework. There is a need to maintain existing performance and to provide fit for purpose services but also there is a need to legislate that companies take into account the sustainability of their assets and to ensure that any further remedial measures to maintain their performance are sustainable.

The Technology Challenge

Local Sustainable Drainage Structures

Future increases in surface runoff, for whatever reason, may overload the existing system and/or result in enhanced environmental damage. CIRIA (2006) proposed a methodology for 'Designing for exceedence in urban drainage' that provides best practice advice for the design and management of urban sewerage and drainage systems to reduce the impacts that arise when flows occur that exceed their capacity. There is therefore a future need to adapt the existing system to accommodate these additional flows. Local adaption is feasible and here reference is not only made to CIRIA but also to the EPSRC and Water Industry funded AUDACIOUS project, managed by Professor Richard Ashley, http://www.eng.brad.ac.uk/

audacious/, that has highlighted the potential for the adoption of Sustainable Drainage Systems at a local level within the catchment.

SUDS are physical structures that are built to receive surface water runoff – ie. the rainfall that runs directly off the catchment surface. Ideally therefore SUDS structures should be positioned as close as possible to where the rainwater falls. SUDS structures provide storage and attenuation of the flow which may subsequently be infiltrated into the ground or safely returned to the drainage system at a time when the system has available capacity. SUDS structures may also provide treatment to improve water quality by the processes of sedimentation, filtration, absorption and biological degradation. There is a range of design options available.

Green roof technology is an emerging sustainable technology in the UK and is currently a topic of much research, see http://www.thegreenroofcentre.co.uk/. In addition, the future house, as shown in Figure 2, will see the implementation of many new infrastructure technologies for water retention and storage, water saving, water recycling and

Figure 2. Water control and re-use local to the house

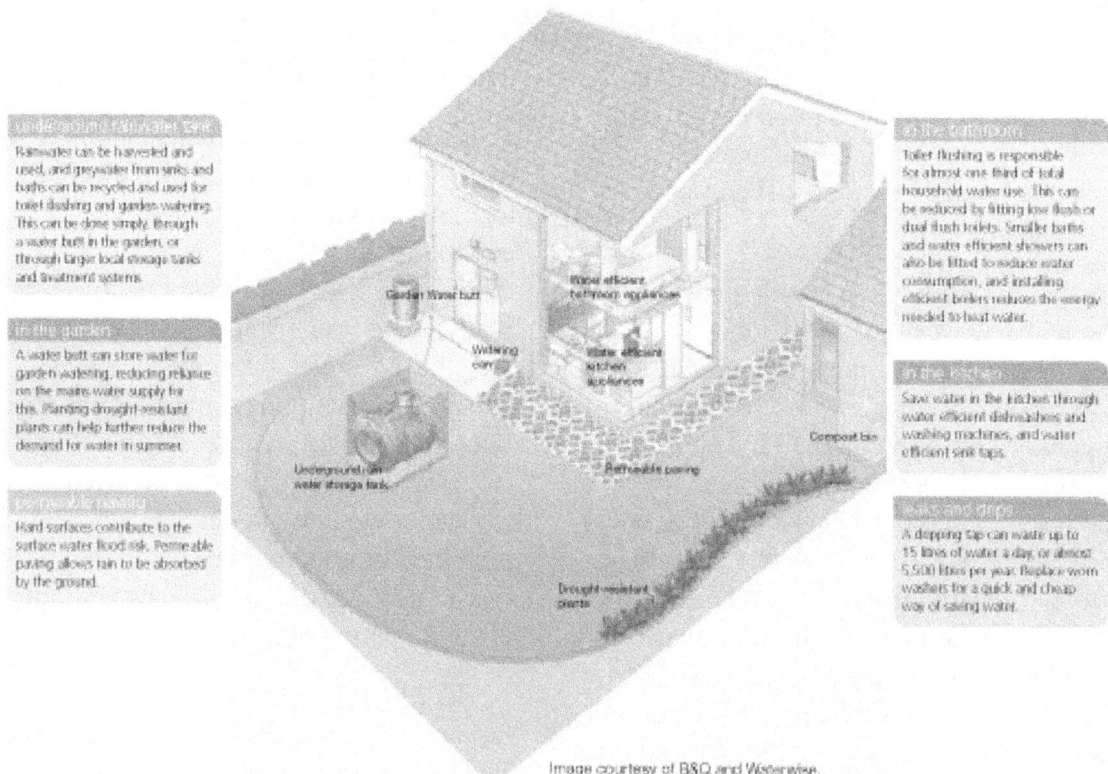

Image courtesy of B&Q and Waterwise.

(Crown Copyright, Figure 7, Future Water)

water re-use. In respect of the management of stormwater that runs off the impervious surfaces close to a property the use of roof run-off water butts and/or disconnection, and below-ground soakaways and storage structures (under drives and lawns), for example, www.hydro-international.biz/index_uk.php and Polypipe www.polypipe.com/polypipe/bp.

These are used either to infiltrate the stormwater into the ground or to lag and attenuate the peak of the stormflow with the stored volume safely returned to the sewerage system at some time after the event, when the system has available capacity. Alternatively the stored water may be re-used. Using such systems it is feasible to achieve water neutrality.

Other sustainable techniques include infiltration trenches and basins, porous pavements, swales, retention and detention ponds and wetland systems. The latest design methodologies, produced as an output from a major EPSRC and Water Industry funded research programme 'Water and New Developments' (WaND), co-ordinated by Professor David Butler of the University of Exeter, may be found at http://www.uksuds.com/ or at CIRIA, http://www.ciria.org/rp752.htm.

SUDS uptake

The uptake of SUDS has been more prevalent in Scotland and elsewhere in Europe when compared to England and Wales, although there is a major project to better understand the implications and impacts of disconnection, currently being undertaken by Welsh Water. The implementation of SUDS in highly urbanised areas is often difficult due to limited access and space, but there are other issues that need to be addressed to improve uptake of these sustainable systems in England. These include:

- Customers have the right to connect into a sewer.

- Service pipes to properties are likely to become the responsibility of the Water Companies who have no authority to implement SUDS on privately owned land.

- There are significant no financial incentives for water companies to deliver sustainable solutions.

- Companies offer only a small rebate of their surface water drainage charges for customers whose surface water is not connected to the public drainage system.

- The current regulatory framework with 5-year AMP periodic reviews does not promote strategic and sustainable solutions, although companies are now required to prepare Surface Water Management Plans and 25-year plans for their sustainable future.

- The promotion of competition between water service providers.

- Changes in legislation and regulation related to the Town Planning and Policy White Paper, the draft Climate Change Bill and the Code for Sustainable Homes and Building Regulations.

As a consequence, the Pitt Review recommends '*Government consulting on options for resolving the barriers to take up of sustainable drainage systems (SUDS), including options for ownership and adoption of these systems across the main agencies involved in urban and land drainage*'. However, lessons learned with the successful implementation of Water Sensitive Urban Design strategies in many Australian cities could provide a major insight into a potential way forward.

Asset Deterioration

It is well recognised that in most UK towns and cities there is an ageing below-ground pipe drainage infrastructure and that these systems continue to deteriorate. In many cases the future may see the rehabilitation and renewal programmes lag behind the rate of deterioration, with a consequent reduction in the level of service. It has to be recognised that some 55% of surface flood events are caused by non pluvial or fluvial events and these are termed flooding 'Other Causes' – due to hydraulic inadequacy, sedimentation, blockage, collapse, Mechanical and Equipment failure, etc. There has been much EPSRC and Industry research to address such surface flooding. For example the EPSRC EA Defra funded Flood Risk Management Research Consortium has researched and developed a statistical model to predict deterioration (Savic, 2006) and the incidence of blockage and collapse and its impact on surface flooding, whilst the industry has developed performance and cost models for individual asset types. It is essential, therefore, that these infrastructure performance models and the way in which they are likely to change in future, form an integral part of any sustainable solution.

Future Awareness

It is important to recognise that the introduction of Sustainable Systems will impact on the performance of existing combined drainage systems and that the introduction of infiltration systems may impact on the local groundwater level and quality. There is a need to be aware or to ensure:

- Existing infrastructure will continue to be used as it is considered uneconomic to abandon it.

- The reduced surface runoff into combined systems should take due regard of the existing available capacity such that the performance of the sustainable system may be optimised, with the proviso that the existing infrastructure and wastewater treatment works do not change.

- Where ground water is low, reductions in stormwater inputs to combined systems may see a 'drying out' of the ground that surrounds the pipes with consequent impacts and soil movements.

- Any increase in groundwater level due to recharge from sustainable systems does not create zones of groundwater flooding.

- Increases in infiltration inflow due to increased groundwater level should not significantly reduce the available capacity of the system.

The required development of Surface Water Management Plans (SWMPs) will see the introduction of new strategies that will result in a paradigm shift in conventional practice. This is evidenced by the current research into urban flood risk as part of FRMRC. From an urban perspective this has attempted to develop a new model that describes the interaction between the performance of the below-ground drainage system and the above-ground surface flows (Djordjevic *et al.*, 2007). What is clear from this study is that it is not feasible for the below-ground system to cope with projected increases in surface run-off due to climate change, urbanisation and population growth and hence **an optimal solution will be to manage the flood flows on the urban surface and to create designated flood flow paths and storage areas**. This will allow for a change in philosophy whereby it is the water level and velocity of flow that are important and not the flow volume. This will identify the need for zones of flood resilient properties.

In respect of new development, SUDS should ideally be constructed on brownfield sites, but new urban planning opportunities could see the introduction of the green grid concept, both for flood alleviation and as heat sinks, and the development of specifically designed flood resilient properties specifically for use on flood plains. New approaches and philosophies may see the introduction of the 'urban flood embankment' at locations where fluvial flooding is encouraged to enter onto the urban area at locations from where the excess surface waters can be safely managed with minimum disruption and impact. For example, to prevent urban flooding in Japan, stormwater detention ponds are constructed within highly urbanised areas and all houses and multi-storey apartments within the pond are constructed on stilts. These detention basins serve as community parks at times of dry weather and hence there is the need for major clean-up operations following each major flood event.

In the UK the opportunity for a paradigm shift in flood risk management will create a massive need for increased public awareness, knowledge and behavioural change, new planning horizons, changes in urban form and infrastructure and the modelling of flood flows over the catchment surface using sophisticated digital elevation and terrain models and complex and interactive 1D (sewer) and 2D (surface) mathematical models. The development of such models is ongoing. Defra's framework for pro-environmental behaviours includes an improved understanding of consumer attitudes and behaviour and the motivations and barriers to individual and community action across a wide range of environmental issues. This will help link water saving to other behaviours on energy, waste, transport and environmentally friendly products.

The adoption of sustainable integrated urban drainage, and the infrastructure that supports it, are therefore key to the management of future flood risk. It is again stressed, therefore, that there is the potential for. 'Urban Flood Risk Maps' to change beyond recognition. Such approaches will afford significant greater protection of our critical infrastructure.

The Future for Infrastructure Operation and Control

Thames Water has recently published 'Taking Care of Water' – a draft 25-year plan for a sustainable future, www.thameswaterconsult. co.uk, and presents an excellent summary of the potential future options for the UK Water Industry, under the umbrella of Customers, Assets and External Environment. In respect of sustainability and infrastructure there is a plethora of potential developments in innovative technologies, remote sensing and communications systems that will allow large quantities of data to be transformed into knowledge. Such information will then be used for the prediction of sustainability indicators such as energy and carbon footprints, and the potential for pollution and flooding incidents. A word of caution, however, is that one water company based their pumping strategy on the wide scale deployment of energy efficient pumps only to find that the carbon footprint associated with their operation and maintenance requirements significantly outweighed, by a factor greater than five, any savings in energy to operate the pumps.

Further sustainable development and change will see improved infrastructure management with a move towards a more proactive operational and maintenance strategy in near real time. Here, Phase 2 of the EPSRC EA/Defra funded FRMRC proposes to develop a near real-time monitoring and **predictive modelling tool for the near real-time forecasting and control of potential urban flood events**. This will provide the opportunity for enhanced warning and mitigation of events. The use of weather radar measurements and nowcasting to predict changes in rainfall at a particular location in future, based on rainfall patterns recorded by radar, together with large numbers of low cost sensors and communications technologies, as shown in Figure 3, will allow the predicted and actual performance to be observed in near real time, thereby providing the option for enhanced control to optimise the sustainable performance of the system. These measures will allow the industry to 'sweat the existing assets' for their optimum sustainability and subsequently to optimise the Capex and Opex spend to maintain the system as sustainable. In the longer term, account may be taken of the impact of adaptation and resilience strategies.

Summary

The future making of our urban drainage systems more sustainable is an exciting and stimulating challenge, and one to which we all look forward.

Figure 3 Strategies for the control of urban drainage infrastructure for optimum sustainable performance

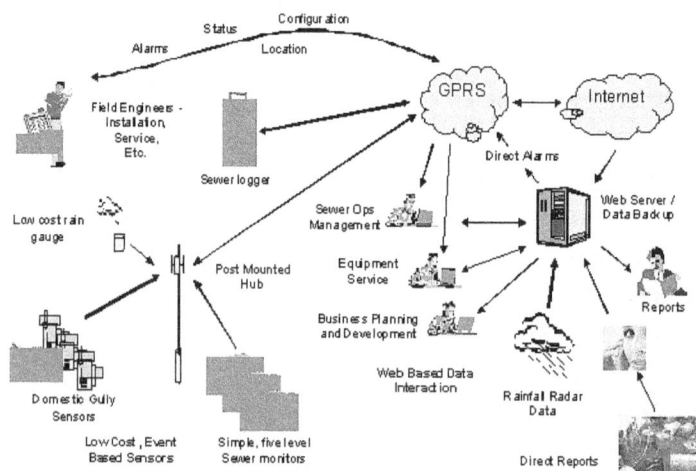

The paper has attempted to highlight a limited exposure to some of the potential ways to implement new and novel approaches to better improve our integrated to approach sustainable drainage.

The future approach to sustainable drainage has to be within an integrated framework. There are a number of barriers to the implementation, primarily institutional, and it will be interesting to see how those responsible for all aspects of urban drainage and surface water management are brought together in harmony. The Government *'propose using surface water management plans as a tool to improve co-ordination between stakeholders involved in surface water drainage, and promoting sustainable drainage systems by clarifying responsibilities and improving incentives for property owners and developers'* (Pitt, 2007).

There are now many sustainable techniques that may be used to manage surface water runoff, *'allowing for the increased capture and reuse of water; slow absorption through the ground; and more above-ground storage and routing of surface water separate from the foul sewer system'* (after Pitt, 2007). The uptake of these systems is more advanced in Scotland when compared to England. Many issues are again Institutional and legislative, but a programme of incentives, to Local Authorities, water service providers, developers and the public, would be of significant benefit.

Awareness of the paradigm shift in the management of most surface water above ground (over and above the capacity of the existing drainage system), will have major implications in terms of public perception, appreciation and acceptance. Public at risk will become a key issue!

Controlling surface water and flood flows with changes in urban form and urban flood pathways could have a significant impact on the location of existing high risk flood areas. Hence there is a need to continually update the impact of new infrastructure with the requirements of the Floods Directive.

The impact of the implementation of this sustainable technology on the performance of the existing drainage system should not be underestimated – it is argued that the existing systems, mainly combined, are here to stay.

New sensors and communications technologies will see advances in performance and optimisation of the new era of sustainable systems and for the control of infrastructure in near real time.

Exciting times!

References

CIRIA (2006) *Designing for exceedance in urban drainage - good practice*. (C635), ISBN (13 digit): 978-0-86017-635-0.

Defra (2004) *Making Space for Water. Developing a new Government strategy for flood and coastal erosion risk management in England*. Department of Environment, Food and Rural Affairs, London.

Defra (2008) *Future Water: The Government's water strategy for England*. The Stationery Office, ISBN 978-0-10-173192-8.

Djordjević, S., Chen, A., Leandro, J., Savić, D., Boonya-aroonnet, S., Maksimović, C., Prodanović, D., Blanksby, J. and Saul, A.J. (2007) *Integrated Sub-Surface/Surface 1D/1D and 1D/2D Modelling of Urban Flooding*. Aquaterra, Amsterdam.

Savic, D., Giustolisi, O., Berardi, L., Shepherd, W., Djordjevic, S. and Saul, A.J. (2006) *Modelling Sewer Failure By Evolutionary Computing*. Proceedings of ICE, Water Management, March.

The Pitt Review (2007) *Learning lessons from the 2007 floods*. Cabinet Office. www.cabinetoffice.gov.uk/thepittreview.

Journal of Practical Ecology and Conservation Volume 7(2) 2009

Frickley Colliery, Wakefield: South Kirby Retention Ponds Case Study

Andy Higham
Wakefield Metropolitan District Council

Introduction

In 2000, an application was submitted for a 4.6ha housing development on the site of a former factory and an adjacent area of undeveloped land in South Elmsall, West Yorkshire.

The site lies between the B6422 to the south west, The Beck (main river) a tributary of the River Don to the north east and to the south east is a drainage ditch (old course of Frickley Beck) carrying run-off from the former Frickley Colliery spoil heaps and surrounding fields (see Figure 1, in appendix). Due to mining subsidence, water in the drainage ditch is now pumped into The Beck by means of a submersible pumping station, which is maintained by Dun Internal Drainage Board (DIDB).

During planning consultation, it was agreed that the developer would be permitted to build within the flood zone, infill and culvert a section of the drainage ditch and be granted a free discharge of surface water to The Beck. The Council agreed on the condition that they mitigate the loss of flood plain, the loss of capacity in the ditch and mitigate the additional surface water run-off.

An accord was made between the developer and Wakefield Metropolitan District Council Land Drainage (WMDC LD) that the council would make available land for an off site scheme and undertake the design & build of such a scheme to meet these conditions, for a sum of £50,000 plus a commuted sum of £70,000 to provide long term maintenance of the scheme.

Objectives of the Scheme

- Mitigate surface water run off from new housing development

- Mitigate development in floodplain

- Mitigate infilling and culverting of minor tributary

- Mitigate any potential impairment of water vole (*Arvicola amphibius*) habitat

- Create aquatic habitats & enhance local ecology

The Scheme

Storage

Consultants were contracted to determine the amount of storage required to mitigate the effects of the development, their findings showed the following amounts of storage were required:

- Flood plain encroachment – $860m^3$

- Additional development run-off – $1040m^3$

- Infilling of drainage ditch – $750m^3$

- Total amount of storage required to mitigate development – $2650m^3$

These figures are based on loss of flood storage due to the development during a flood event of a 100-year return period.

Site

An appropriate location for a retention facility was agreed to be on the Frickley Colliery reclamation site, upstream of the development adjacent to Frickley Beck a tributary of The Beck (see Figure 1). An area of approximately 1.5ha already susceptible to flooding which is presently being used to graze horses, was chosen (see Figure 2).

An ecological survey of the site showed that the majority of the site comprised of a rank sward of tussocky grassland with some scrub made up of hawthorn (*Crataeus monogyna*), willow (*Salix* sp.) *etc* and dense

Figure 1. Location of Site

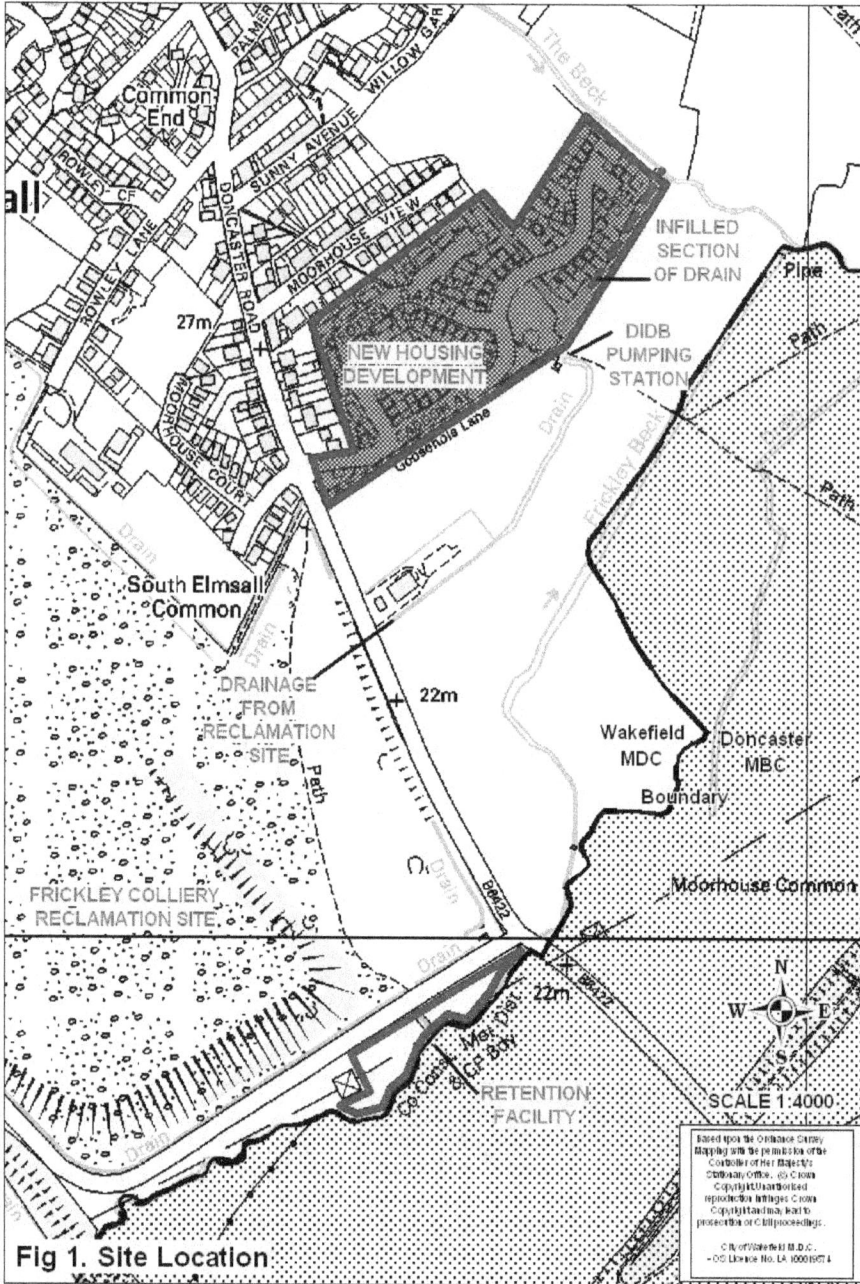

Fig 1. Site Location

Figure 2. Site prior to works, May 2003

thickets of bramble (*Rubus futicosus*). Two small areas of wet grassland dominated by creeping bent (*Agrostis arvense*), hard rush (*Juncus inflexus*) and jointed rush (*Juncus articuatus*) were identified as the main areas of interest and although no protected species were found the report suggested these areas be retained. There was some evidence of water vole activity in Frickley Beck and the report stated that the scheme was required to have no adverse affect on water vole habitat.

Rather than adversely affecting ecology, the report viewed the scheme as an opportunity to increase biodiversity and improve wildlife habitats.

Design & Build

Initially the developers commissioned a design for the scheme but the outcome was a highly engineered design, so WMDC LD modified the design to produce a softer plan and it was this that was carried forward. This design was devised to create a maximum of 2800 – 3000m^3 of storage.

WMDC LD drew up and sent out to tender a contract for the scheme and a contractor was then selected. The majority of the works were started and completed between February and March 2004, at a contract cost of around £42, 000.

The general layout of the scheme can be seen in Figure 3, the storage is provided by two separate ponds, located either side of the existing settlement channel which accepts potentially contaminated run-off flows from the Frickley Colliery reclamation site before issuing into Frickley Beck.

Royal Society for the Prevention of Accidents (RSPA) guidelines were built-in to the design, including safety shelving and some buffer planting.

The pond design incorporated ecological enhancement, with bank slopes varying between 1:3 and 1:7. Other steeper sides were dressed with weaved willow to create habitat for water voles and here an intermediate safety shelving profile was used. The beds of the ponds were designed to have a patchwork of hummocks and hollows within the draw

down zone. Both ponds and areas within them have differing depths of water to attract a diversity of flora & fauna. Bank profiles were varied to provide narrow inlets and wide bays.

The ponds were planted with locally sourced marginal and wetland plants and the surroundings sowed with a wildflower mixture, to provide some immediate interest and to help suppress unwanted plants.

It was agreed that the excavated material from the ponds would be incorporated into the Frickley Colliery reclamation site.

Maintenance & Management

Maintenance of the ponds has been carried out several times since completion of the scheme. However, neither an ecological management plan nor hydrologic maintenance regime has been produced. Currently maintenance, such as removal of debris to preserve the hydrological function of the scheme, is carried out on a very ad hoc basis. At the same time, reduction of invasive species such as reedmace (*Typha latifolia*) has been carried out, again to preserve the hydrological function of the pond and to a lesser extent to help maintain the diversity of flora and fauna.

The ponds are now being used as a source of marginal and aquatic plants for colonisation of other WMDC LD Schemes.

Discussions & Conclusions

The Frickley retention ponds are providing flood storage, which mitigate the loss of flood plain storage and additional run-off caused by the housing development.

Nevertheless, the scheme provides little additional storage. This scheme is just maintaining the status quo, very little storage has been provided to reduce existing flooding which already occurred downstream in the catchment.

Future schemes in the catchment have the potential to create additional flood protection but this may require collaboration between WMDC and Doncaster Metropolitan Borough Council and will also be subject to funding issues.

Figure 3 Plan of Reclamation Scheme

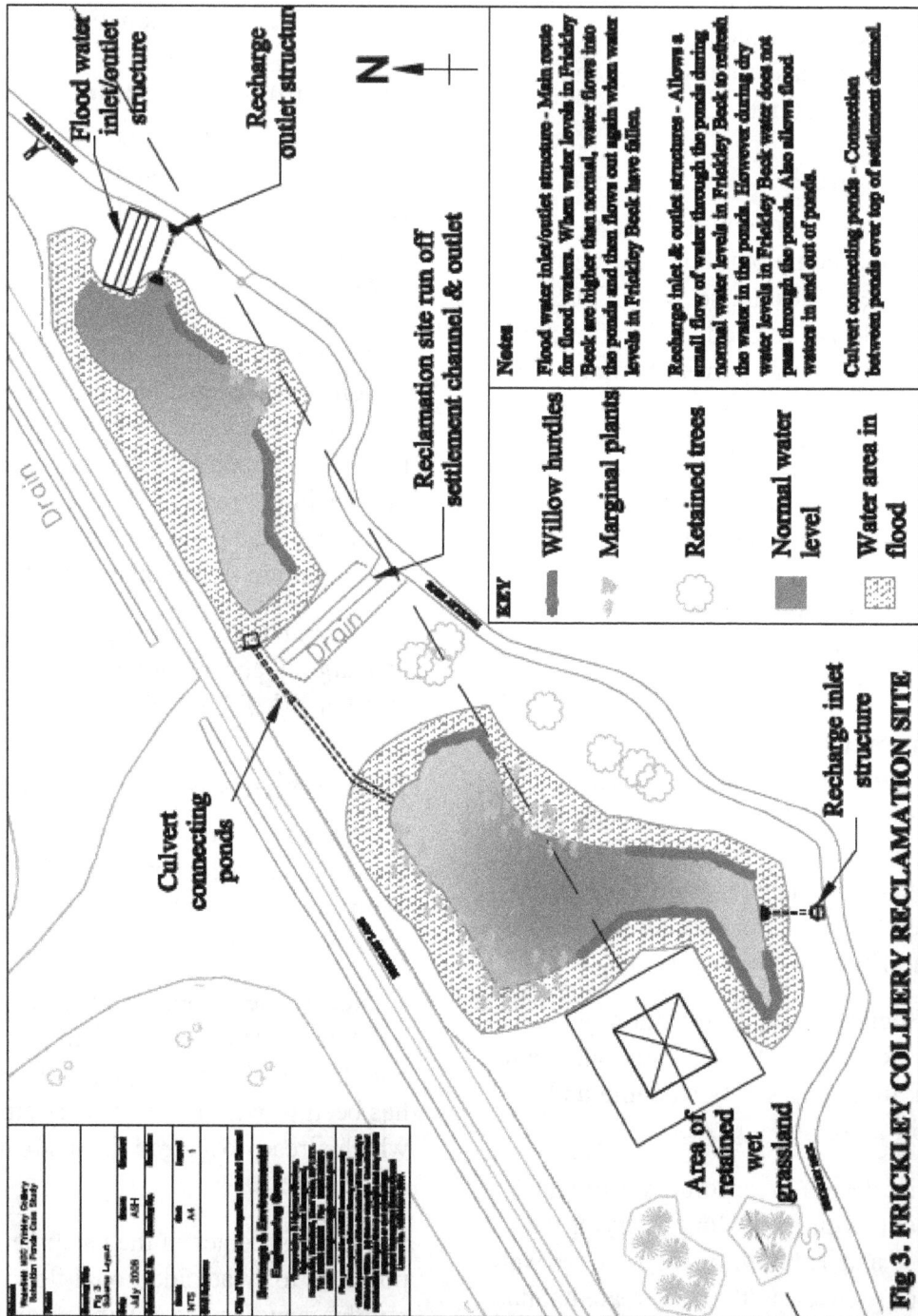

Notes

Flood water inlet/outlet structure - Main route for flood waters. When water levels in Frickley Beck are higher than normal, water flows into the ponds and then flows out again when water levels in Frickley Beck have fallen.

Recharge inlet & outlet structures - Allows a small flow of water through the ponds during normal water levels in Frickley Beck to refresh the water in the ponds. However during dry water levels in Frickley Beck water does not pass through the ponds. Also allows flood waters in and out of ponds.

Culvert connecting ponds - Connection between ponds over top of settlement channel

KEY

Willow hurdles

Marginal plants

Retained trees

Normal water level

Water area in flood

Fig 3. FRICKLEY COLLIERY RECLAMATION SITE

Figures 4 to 7 show the ponds at several stages during their construction

Figure 4. Willow hurdles under construction

Figure 5. Pond B during works, February 2004

Figure 6. Pond B post works, May 2004

Figure 7. Pond B, October 2006

Journal of Practical Ecology and Conservation Volume 7(2) 2009

In June 2007 during a period of heavy rain, which affected several areas in the WMDC area, a number of dwellings on the new development suffered from flooding. Was this due to the development site not being moved above the hundred-year flood zone? Or was the flooding event greater than the hundred-year return period?

A great advantage was gained by incorporating excavated material into the reclamation site. Usually excavated material has to be kept on location and so potential flood storage areas are taken up with disposal of excavated material or a great expense is incurred in removal of material off-site.

As an alternative to traditional flood storage measures on development, e.g. large storage tanks underground, the ponds have enhanced the local ecology. They have turned land of very low ecological value into an area with moderate biodiversity value, with good potential to become more valuable to biodiversity in the future.

Should the Frickley Ponds of been planted in the first place and should we be using them as a source of plants? The Ponds Conservation Trust (2001) recommends that new ponds are not planted but allowed to colonise and mature naturally. They state reasons for this which include that new ponds are a very distinctive habitat, used by plants and animals that are not found in mature ponds, and also that there is an increased risk of accidental transfer of invasive alien species when planting. This obviously has further implications due to the pond being used as a source of plants to colonise other Wakefield Metropolitan District Council LD schemes.

Future Projects & Considerations

- Development and delivery of a maintenance programme to maintain the scheme's hydrological function.

- Development and delivery of an ecological management programme alongside regular ecological surveys to preserve and improve the scheme's biodiversity.

- Improve understanding of why the housing development flooded in June 2007.

- Creation of additional flood storage facilities to provide additional flood storage within the catchment.

Figure 8. Pond B, autumn 2007

Other Schemes

This case study is not a report on a demonstration site. Frickley colliery retention ponds are just part of the WMDC LD section's district wide programme of flood mitigation schemes for both new development and existing flood problems. Below is a list and brief outline of just some of the other schemes created by WMDC LD.

Southern Washlands Nature Reserve, River Calder, Wakefield

- Formerly a gravel extraction works and landfill site, the area was reclaimed in 1988 using funding from the Department of the Environment (government body). Also funding was granted by Yorkshire Water Rivers Division (now part of the Environment Agency) to create controlled washlands within the site.

- Created to mitigate the infilling of uncontrolled washland downstream on Welbeck landfill site.

- Now a local nature reserve the site comprises a variety of habitats including swamp and marshy grassland and has good public access.

- Flood storage provided – c. 38,0000 m^3

- Engineering and landscape costs – c. £250,000.

Figure 9. Southern Washlands Nature Reserve, River Calder, Wakefield

Journal of Practical Ecology and Conservation Volume 7(2) 2009

Lupset Golf Course

- Created in 2003 this developer funded retention pond mitigates both run off from housing development and loss of Great Crested Newt (*Triturus cristatus*) habitat.

- Created a feature on a WMDC run public golf course.

- Flood storage provided – *c.* 100 m^3

- Engineering & landscape costs *c.* £4,000.

Willowbridge Beck, Whitwood Golf Course

- Created in 2006 on a WMDC public golf course, this developer-funded scheme has increased the watercourses' flood plain to mitigate run off from an industrial development up stream.

- Enhanced habitat for Water Vole (*Arvicola amphibius*) and improved overall biodiversity of a 200m stretch of watercourse.

- Flood storage provided – *c.* 1,600 m^3

- Engineering and landscape costs *c.* £30,000.

Figure 10. Lupset Golf Course

Figure 11. Willowbridge Beck, Whitwood Golf Course

Langthwaite Lane, Moorthorpe

- Built on WMDC-owned land this recently completed developer and capital funded retention pond mitigates several developments upstream and reduces existing/historical flooding to a number of dwellings and businesses downstream in South Elmsall.

- Future potential to become a very ecologically diverse wetland habitat.

- Flood storage provided – *c.* 7,500 m^3

- Engineering and landscape costs – *c.* £80,000.

St Johns Pond, Normanton

- Created in 1999 on WMDC-owned land, this developer funded retention pond and treatment reed bed mitigates run off from a housing development upstream and provides extra flood storage in the catchment.

- Promotes a diverse mix of wet, damp and dry habitats.

- Flood storage provided – *c.* 2,500 m^3

- Engineering and landscape costs – *c.* £80,000.

Figure 12. Langthwaite Lane, Moorthorpe

Figure 13. St Johns Pond, Normanton

Oakenshaw Beck, Agbrigg

Areas of Agbrigg have flooded several times in recent years and in June 2007 approximately 200 dwellings and businesses were flooded. Consequently, Wakefield Metropolitan District Council LD created a scheme on WMDC-owned land with works due to start late 2008. Designed to increase flood protection to a large area of Agbrigg, the scheme involved improved control of floodwaters as well as increasing flood storage by a further *c.* 40,000 square metres in the flood plain, at a cost of approximately £200,000. Ecologists and local biodiversity groups have been consulted to ensure that the scheme will also provide good aquatic and wetland habitats.

References

The Ponds Conservation Trust (2001) *Planting-up Ponds* (PCT REF:2001/08).

Water quality in a sample of wells in the Al Jebel Al-Akhader region of north-eastern Libya and the implications for vegetation

Younis Abdoul Moula Alhendawi[1], Mohamed Kamel Ahmed[2] and Ian D. Rotherham[3]

[1] Faculty of Natural Resources, University of Omar Almkhanar, Tobruk, Libya.
  alhendawiy@hotmail.com

[2] Department of Botany, Faculty of Sciences, South Valley University, 83523 Qena, Egypt.
  mohamed-kamel@lycos.com

[3] Sheffield Hallam University, UK.
  i.d.rotherham@shu.ac.uk

Abstract

Samples of water were collected from fourteen wells in Al Jebel Al-Akhader, north-eastern Libya. Seven wells were from the northern side, which faces the Mediterranean Sea; and the other seven wells were on the southern side, facing the Sahara Desert. Total dissolved salts TDS, pH, electrical conductivity (EC) and selected dissolved ions (Na^+, K^+, Ca^{2+}, Mg^{2+}, Cl^-, HCO_3^- and SO_4^{2-}) were estimated. All parameters were elevated in the samples collected from the southern side in comparison with those from the north-east. In particular, salinity was increased in the southern side samples. The vegetation on the northern side was dense and had relatively high species diversity, but southern side was dominated by sparse vegetation of a limited range of desert plants. It was concluded that in the southern area the high water salinity together with high temperatures and low rainfall influenced the vegetation structure. The latter was poor and sparse.

Key words: ground water, Jebel Al-Akhader, salinity, water quality, desert vegetation.

Introduction

Water is indispensable for all forms of life and is needed for many human activities. Access to safe freshwater is now regarded as a universal human right (United Nations Committee on Economic, Social and Cultural Rights, 2003), and the Millennium Development Goals include the extended access to safe drinking water and sanitation (United Nations Development Programme [UNDP], 2006). Sustainable management of freshwater resources has gained importance at regional (e.g. European Union, 2000) and global scales (United Nations, 2002, 2006; World Water Council, 2006), and 'Integrated Water Resources Management' has become the corresponding scientific paradigm.

Climate is a key factor determining different characteristics and distributions of natural and managed systems, including hydrology and water resources. Many aspects of climate influence various characteristics and distributions of physical and biological systems. These include temperature and precipitation, and their variability on all timescales from days to the seasonal cycle to inter-annual variations.

Groundwater in shallow aquifers is part of the hydrological cycle and is affected by climate variability and change through recharge processes (Chen *et al.*, 2002), as well as by human interventions in many locations (Petheram *et al.*, 2001). In the Upper Carbonate Aquifer near Winnipeg, Canada, shallow well hydrographs show no obvious trends, but exhibit variations of three to four years correlated with changes in annual temperature and precipitation (Ferguson and George, 2003). Groundwater systems generally respond more slowly to climate change than surface water systems. Groundwater levels correlate more strongly with precipitation than with temperature, but temperature becomes more important for shallow aquifers and in warm periods.

Water deficiency is one of the greatest problems which face human society on planet earth. The need for water increases with population growth and so water deficiency will be an increasing issue especially in the Middle East and North Africa (Kamel, 2002). The demand for groundwater is likely to increase in the future, the main reason being increased water use globally.

One of the interesting topographical features in Libya is Jebel Al-Akhader, which lies in north-eastern Libya and extends east-west. So, the northern side faces the Mediterranean Sea whilst the southern side faces the Sahara Desert. Jebel Al-Akhader is a heavily forested, fertile upland area; one of the few forested areas of Libya which is one of the least forested countries in the world. Jebel Al-Akhader has an average elevation of 500-600m (masl), with a maximum of 880m (masl) in Sidi Al-Hummry (El-Zwam, 1995). The main water resources for the population are the winter rains, wells and springs.

The current investigation complements the studies by Al-Hendawi (2004) and by Al-Hendawi and Kamel (2007). The aim of this work is to establish scientific data in terms of chemical and physical characteristics of the water resources of Jabal Al-Akhader basin in the north-eastern region of Libya. The aim was also to analyse water quality of the north-east region of Libya, particularly of agricultural land.

Materials and Methods

Study area:

The study area as shown in (Figure 1) is located in the Al Jabal Al Akhdar region. This is a mountain range along the northern coast of north-eastern Libya, located approximately 31° N ,and 23° E ,with aerial extent about 20,000 km² from Wadi Al Bab in the west to Tamimi in the east, and from Taknis and Sidi Al Humree in the north to the Balat areas in the south. The climate in Al Jebel Al-Akhader region varies according to the geographical distribution of the area. The climate of the south Al Jebel Al-Akhader is classified as 'semi-arid' in the northern parts of the study area, and 'arid' in the south. According to

Figure 1. Map to indicate the study location of Jebel Al-Akhader in Libya

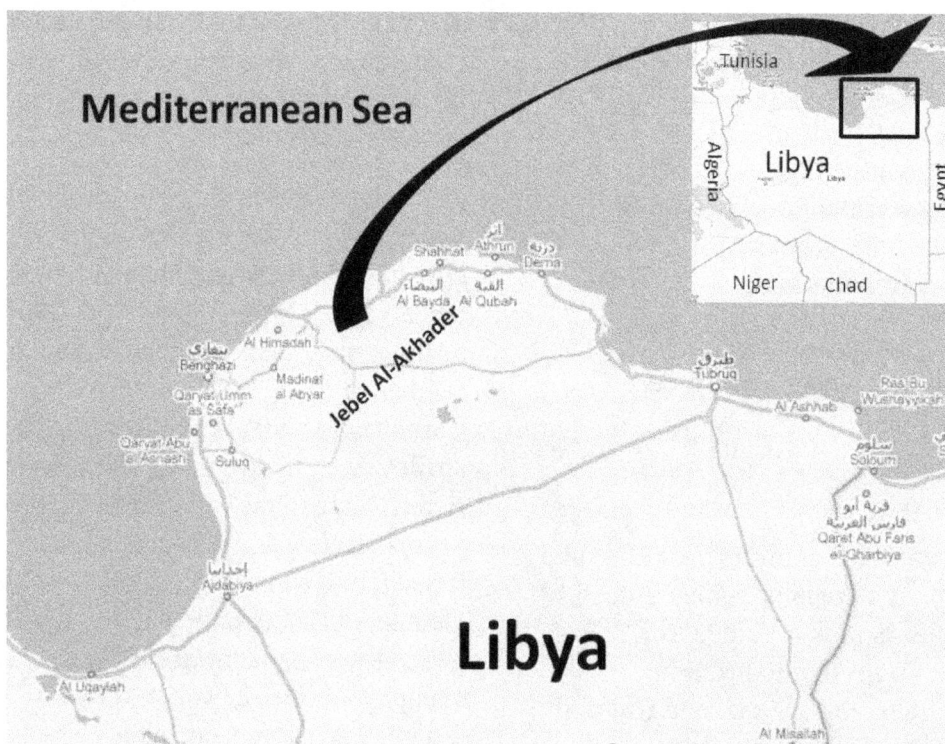

(SJP, 2005) the weather is as follows. The rainy season is from October to March and April, but the distribution and amount of rain show great differences between years. The average annual precipitation in the survey area is about 300 mm in the north, and less than 50 mm in the south-east. The annual mean temperature is about 20°C. The coldest period is from December to February with minimum temperatures close to 0°C. The mean January temperature is 8-12°C. The highest temperatures occur from June to August with maximum temperatures over 40°C. The mean July temperature is 22-26°C. The most important winds are the Mediterranean winds from the west bringing the rainstorms and the dry and hot Qibli wind from southern desert areas. Most of the south Al Jebel Al-Akhader region has a desert climate. Only a small area in the north-west has a semi-arid steppe climate. Relative humidity in the northern parts of the study area is from 65% to 75 % during spring and summer time, and up to 90% in winter. In the southern parts of the area, relative humidity is from 40% to 60% during spring and summer, and 75% in winter. Evaporation is very high with about 7-10 mm/day.

Six samples were taken at different times from each well in both areas North and South of the Jebel Al-Akhader. The Northern area sites were: Marawah 1, Marawah 2, Salantah E, Abid E.W, Qandulah, Taknis W and Taknis S wells. The Southern area included the following regions: Mkhili, Al Izziyat, Martubah, Umm Ar-Razam, Tahimi, Kharrubah and Saluq. Seven wells were selected from the southern area, which are indicated by the following numbers: WD1S-2 (D1-2); W47-41 (S1-41); W48-41(S1-41); Wb1w-14 (B1-14); W13-41 (H1-41); W14-41 (H1-41); and W18-41(G1-41).

Six water sample replicates were collected in clean sterilised glass containers at different times during the day and the average values were recorded. The samples were protected from sunlight and transferred quickly to the laboratory. The samples temperature was taken directly at the Location, using a Mercury thermometer. Total dissolved salts (TDS g/1) were measured by Total Solids

Dried at 103 - 105°C as described by Symons and Morey (1941). Electrical conductivity (^mho/cm) was measured using a Conductivity Meter, Model Aol-10 (Dkk Corroration) as described by Arnold *et al.*, 1992). The value of pH was measured using a pH meter as described by Arnold *et al.* (1992).

Chloride (meq/1) was estimated according to Kolthoff and Stenger, 1947. Sulphate (mg/1) was estimated gravimetrically as described by Hillebrand *et al.* (1953. Nitrate (meq/1) was estimated using the Electrode Method as described by (Langmuir and Jacobson, 1970). Calcium and Magnesium (mg/1) were estimated using the titration with EDTA method, U.S. Salinity Laboratory 1954, U.S.D.A. Handbook. Sodium and Potassium concentrations (mg/1) were determined by using the Flame Emission Photometer Method (Barnes *et al.*, 1945). Some samples have to be diluted in pure water at a ratio of 1/100. Molar ratios of various important ionic species were then calculated as follows: Ca/Mg; SO_4/Cl; Na/Cl; and sodium adsorption ratio (SAR). Sodium Absorption Ratio (SAR) was calculated as:

$$ SAR = \frac{Na}{\sqrt{\dfrac{Ca+Mg}{2}}} $$

There were two field trips to survey and assess the vegetation in the northern and southern areas of Al Jebel Al-Akhader as described earlier. Plants were collected and identified according to Ali and Jafri (1977), Jafri and El-Gadi (1986) and El-Gadi (1989).

Results and Discussion

As shown in Figure 2 the pH values were around 7.5–8.0 with the tendency to alkalinity due to the structure of the soil that derives from degraded limestone rocks. The regular rainfall on the northern side results in leaching of cations that are then found in the ground water (Teng-Chiu *et al.*, 2001), but the richness of soil with calcium compensates for the leaching effect. On the other side, the low and irregular rainfall combined with high evaporation, draws cations up the soil profile. Partel (2002) argues that the nature of the relationship between plant species density and soil pH is a function of the composition of species pools, which are affected by the commonness of soils with different acidity (Taylor *et al.*, 1990; Zobel, 1992, Partel *et al.*, 1996; Zobel *et al.*, 1998). Partel (2002) also showed that positive relationships between plant species density and soil pH occur in temperate areas where the evolutionary centres have been in areas with high soil pH.

In this case study however, it seems the over-riding influence on vegetation is not pH directly, but salinity and drought.

With the decreasing rainfall southward, the salinity and, consequently, electrical conductivity increased (Figure 3). While the TDS ranged between 0.5–1.0 g/l in the northern side of Jebel Al-Akhader, the lowest value was 3.5 g/l in the southern side [WD1S-2 (D1-2)] and increased to 28.6 g/l in the well [W18-41 (G1-41)]. This result accords with reported from the Council of Agricultural Development in 1974. Flores *et*

Figure 2. The estimated pH, total dissolved salts (g/l) and electrical conductivity (μ Siemens / cm at 20°C).

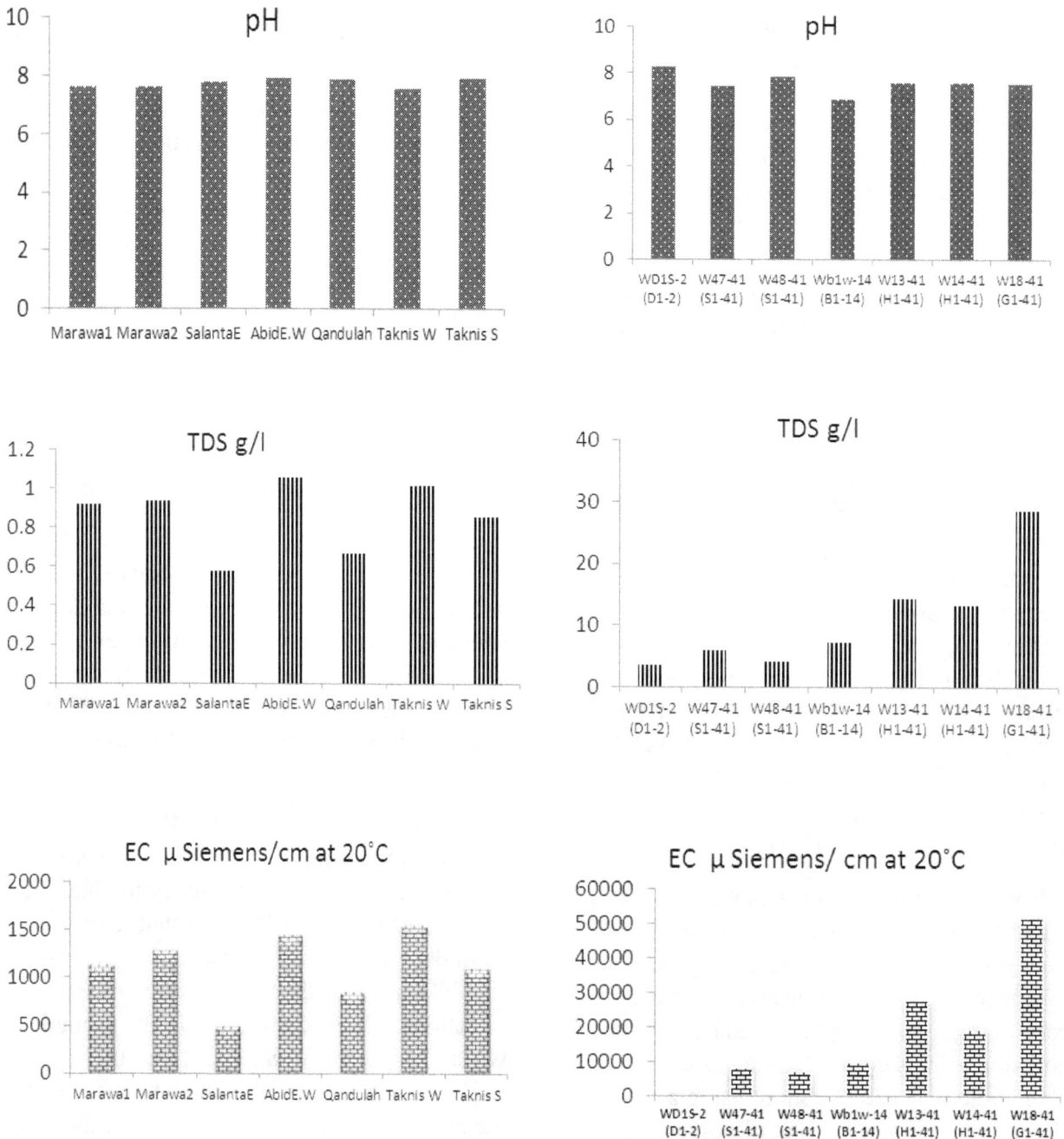

al. (2001) found salinity exposure affects transport processes in plants and alters both nutritional status and tissue ion balance; the result being lowered plant growth.

The increase in salinity due to low rainfall and high evaporation rate affected the water quality and both the nature and density of vegetation cover. For example, in the southern side of the Jebel Al-Akhader region, from the high ground where annual rainfall is about 400 mm (subhumid zone) to the Kharrubah-Mkhili lowlands with annual rainfall from 50 to 100 mm (arid zone), the vegetation cover changes considerably. Indeed over less than 50 kilometres forest changes to desert vegetation. Dense vegetation of the following species dominates the northern side: *Cupressus sempevirens, Pinus halepensis, Juniperus phoenica, Quercus coccifera, Olea europaea, Ceratonia siliqua, Arbutu pavary,* and *Pitacia lentiscus* (Johnson, 1973). On the other hand, the lower lands are dominated by desert plants as *Tamarix, Pulicaria, Ephedra, Silene, Capparis, Zygophyllum,* and *Zilla* species.

Figure 3. The estimated bicarbonate. Sulphate and chloride anions (meq/l) in the water samples collected from studied locations.

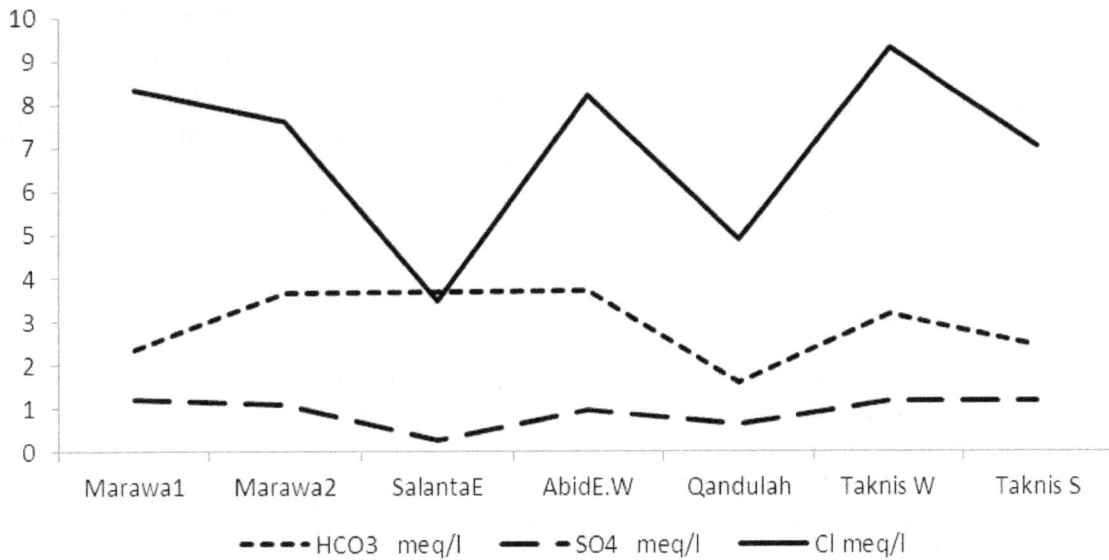

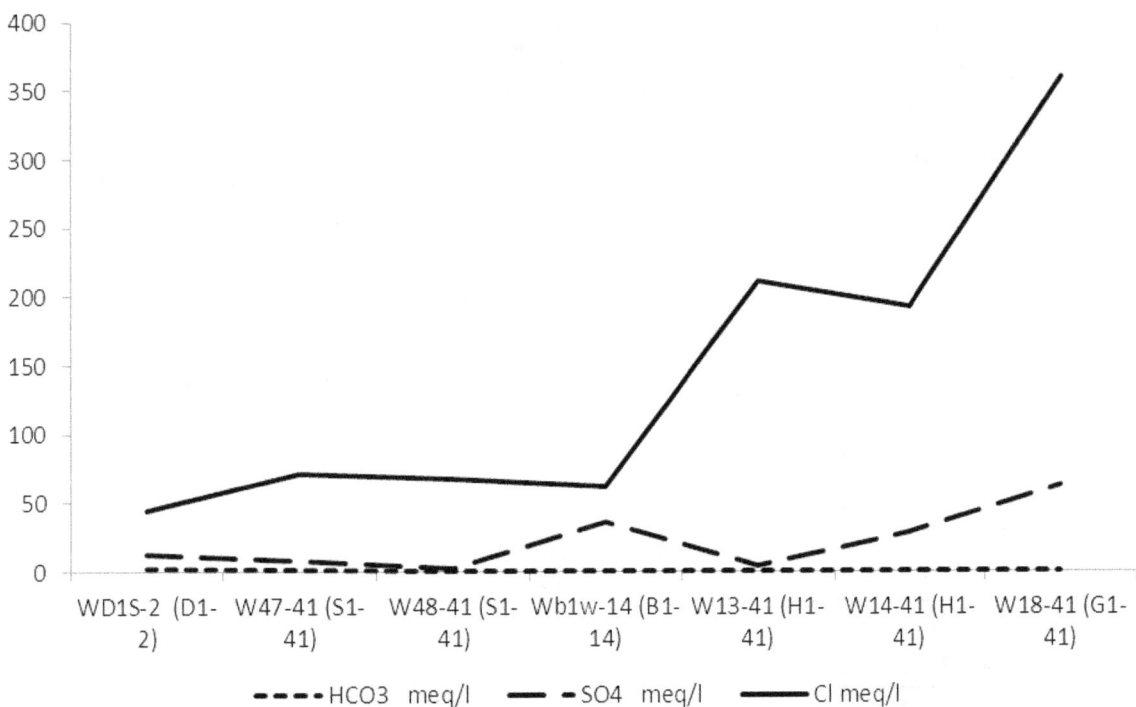

Both estimated cations and anions increased in the water samples collected from the southern side. The dominant cation in the samples collected from northern and southern sides was sodium (Figure 4). Under excessively saline conditions plants take up high amounts of Na^+ while uptake of K^+ and Ca^{2+} are significantly reduced (Ashraf *et al.*, 2004). The dominant anion was chloride (Figure 4). The concentration of sodium ranged between 2.5–6.5 meq in the northern side. On the southern side, the concentration increased and ranged from 34 meq to 289 meq. Chlorides showed low concentrations in the samples collected from the northern side (4 to 10 meq). The concentration of Cl^- in southern side samples was very high (45 – 360 meq). The computed Na/Cl value (Tables 1, 2) indicated that the chloride concentration exceeding sodium in both areas.

Soil is affected by salinity and this may be expressed as SAR, an index representing the ratio between sodium (Na), and calcium and magnesium (Ca & Mg) in water. Our data showed high SAR ratios in the water samples collected from the southern side of Al Jebel Al-Akhader (Table 2). The values ranged between 9 and 45. The major immediate hazard of irrigation with high SAR ratio (SAR>6, EC<4 ds/m) is sodifying the soil surface and deep soil layers. During irrigation cycles, sodium will replace calcium on the soil clay particles and, destabilize the soil structure. This may, and often does, reduce soil surface infiltration to water and drainage of the root zone through lower layers (Nava *et al.*, 2000). The values of SAR were between 2 and 4.

Calcium and magnesium concentrations were high in samples collected from the southern side (Figure 4), ranging from 8–53 meq and 10–84 meq respectively. From the northern side, Ca and Mg concentrations varied between 2–6 meq and 1.5–4.5 meq respectively. The high content of both Ca and Mg in comparison with the sodium content in the northern side decreased the SAR. The decreased soil sodification led to stabilization of soil structure, and decreased sodium competition against calcium and magnesium

Table 1. The computed sodium adsorption ratio, Mg/Ca, SO_4/Cl and Na/Cl ratios of the water samples collected from the Northern flank of the Jebel Al-Akahder area.

	SAR	Mg / Ca	SO_4 / Cl	Na / Cl
Marawa1	2.30	1.90	0.22	0.51
Marawa2	2.40	1.02	0.14	0.63
SalantaE	1.48	1.00	0.07	0.67
AbidE.W	2.73	0.29	0.12	0.65
Qandulah	2.16	1.14	0.13	0.75
Taknis W	4.08	0.79	0.12	0.70
Taknis S	3.71	0.57	0.16	0.87

Table 2. The computed sodium adsorption ratio, Mg/Ca, SO_4/Cl and Na/Cl ratios of the water samples collected from the Souther flank of the Jebel Al-Akhader area.

	SAR	Mg / Ca	SO_4 / Cl	Na / Cl
WD1S-2(D1-2)	09.88	0.72	0.27	0.76
W47-41 (S1-41)	16.10	1.99	0.12	0.80
W48-41 (S1-41)	14.18	1.53	0.03	0.71
Wb1w-14 (B1-14)	09.09	0.62	0.58	0.75
W13-41 (H1-41)	44.97	3.15	0.02	0.91
W14-41 (H1-41)	28.31	0.23	0.15	0.84
W18-41 (G1-41)	35.18	1.63	0.18	0.80

Figure 4. The concentration of calcium, magnesium, sodium and potassium (meq/l) in the collected samples of water from different locations.

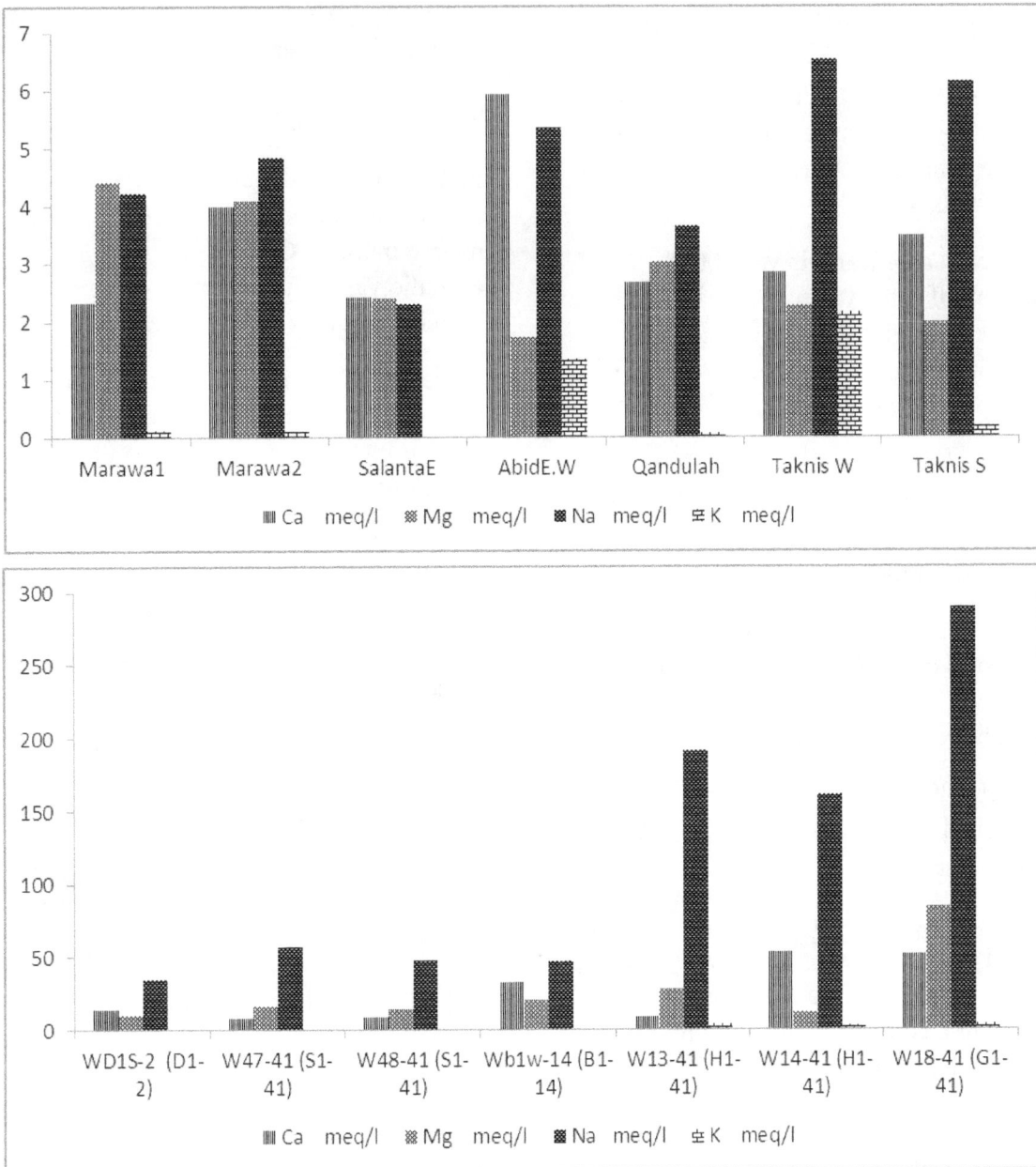

in plant nutritional uptake of cations. We suggest that this may be advantageous in terms of encouraging higher plant diversity in the northern area of Al Jebel Al-Akhader, and is combined with lower temperatures and higher rainfall when compared with the more southerly lowland areas. The proportionally higher concentrations of calcium and magnesium comparison with the total estimated solutes in soils from the northern area resulted in decreased sulphate content. This may be due to the formation of calcium and magnesium sulphates, which are relatively insoluble in water. The soil in the northern area originates from limestone and so

bicarbonate levels are higher from the water samples collected from southern side (Tables 1 & 2).

Conclusions

The overall conclusion is that both climate and location affected ground water quality, and both influence the local vegetation characteristics. In particular, the observed differences between the vegetation type (and its diversity) and the plant cover are affected by the climatic differences directly in the northern and southern areas, and by the influences of ground water quality as well. The latter is in part a result of climatic variations.

39

References

Al-Hendawi, Y. (2004) *A comparative Analysis of Water Quality in Two Regions in Libya to Assess Current and Future Water Problems in NE Libya*. Unpublished PhD. thesis, School of Archaeological, Geological and Environmental Sciences, University of Bradford, UK.

Al-Hendawi, Y. & Kamel, M. (2007) Assessment of three water wellheads as sources of water necessary for human activities in Jabal Al-Akhder area (Libya). *Assiut Univ. Journal of Botany*, **36(2)**, 199-212.

Ali, S.I. and Jafri, S.M.H. (1977) *Flora of Libya. Vol. 1-24*, Department of Botany, Al-Faateh Univ., Tripoli, Libya.

Arnold, E.G., Lenore, S.C. and Andrew, E.D. (1992) *Standard Methods for the Examination of Water and Waste Water*. 18th ed. American Public Health Associations, USA.

Ashraf, M. (2004) Some Important Physiological Selection Criteria for Salt Tolerance in Plants. *Flora*. **199**, 361-377.

Barnes, R.B, Richardson, D., Berry, J.W. and Hood, R.L. (1945) *Flame photometry a rapid analytical procedure*. Industrial and Engineering Chemistry, Washington: American Chemical Society , v. 17, n. 10, p. 605.

Chen, Z., Grasby, S.E. and Osadetz, K.G. (2002) Predicting average annual groundwater levels from climatic variables: an empirical model. *Journal of Hydrology*, **260**, 102-117.

Council of Agricultural Development Report, (1974). *Water Resources Study of the Southern flank of the Jebel Al-Akhdar*. Unpublished Report.

El-Gadi, A. (1989) *Flora of Libya. Vol. 145-147*, Department of Botany, Al-Faateh Univ., Tripoli, Libya.

El-Zwaam, S.M. (1995) *El-Jabal El-Akhdar* (Arabic ed.). Garyounis University, Benghazi, Libya.

European Union (2000) EU Water Framework Directive: Directive 2000/60/EC of the European Parliament and of the Council establishing a framework for the Community action in the field of water policy. *EU Official Journal* (OJ L 327, 22 December 2000).

Ferguson, G. and George, S.S. (2003) Historical and estimated ground water levels near Winnipeg, Canada and their sensitivity to climatic variability. *Journal of American Water Resources Association*, **39**, 1249-1259.

Flores, P., Carvajal, M., Cerda, A. and Marinez, V. (2001) Salinity and Ammonium/Nitrate Interactions on Tomato Plant Development, Nutrition and Metabolites. *Journal of Plant Nutrition*, **24**, 1561-1573.

Hillebrand, W.F., Lundell, G.E.F., Bright, H.A. and Hoffman, J.I. (1953) *Applied Inorganic Analysis*. (Second Ed) Wiley, New York.

Jafri, S.M.H. and El-Gadi, A. (1986) *Flora of Libya. Vol. 25-144*, Department of Botany, Al-Faateh Univ., Tripoli, Libya.

Johnson, D.L. (1973) *Jabal Al-Akhdar, Cyrenaica*. University of Chicago, Illinois, USA.

Kamel, M. (2002) The effect of sudden sodium chloride stress on the ion composition and mechanism of osmotic adjustment in *Vicia faba*. *Pakistan Journal of Biological Sciences*, **5**, 885 -890.

Kolthoff, I.M. & Stenger, V.A. (1947) *Volumetric Analysis. Second edition. Vol. II: Titration Methods: Acid–Base, Precipitation and Complex Reactions*. Interscience Publishers Inc., New York.

Langmuir, D. & Jacobson, R.I. (1970) Specific ion electrode determination of nitrate in some fresh waters and sewage effluents. *Environmental Science and Technology*, **4**, 835.

Nava, H., Bachmat, Y., Shalhevet, S. & Yaron. D. (2000) *Effect of urban development on water quality environmental concerns*. The 40th Congress of the European Regional Science Association European Monetary Union and Regional Policy. Barcelona.

Partel, M. (2002) Local plant diversity patterns and evolutionary history at the regional scale. *Ecology*, **83**, 2361-2366.

Partel, M., Zobel, M. Zobel, K. and van der Maarel, E. (1996) The species pool and its relation to species richness: evidence from Estonian plant communities. *Oikos*, **75**, 111-117.

Petheram, C., Walker, G. Grayson, R. Thierfelder, T. and Zhang, L. (2001) Towards a framework for predicting impacts of land-use on recharge. *Australian Journal of Soil Research*, **40**, 397-417.

Taylor, D.R., Aarssen, L.W. and Loehle, C. (1990) On the relationship between r/K selection and environmental carrying-capacity: a new habitat template for plant life-history strategies. *Oikos*, **58**, 239-250.

Teng-Chiu Lin, Hamburg, S.P., Yue-Joe Hsia, Hen-Biau King, Lih-Jih Wang, and Kuo-Chuan Lin (2001) Base cation leaching from the canopy of a subtropical rainforest in northeastern Taiwan. *Canadian Journal of Forest Research*, **31(7)**, 1156–1163.

United Nations (2002) *Johannesburg Plan of Implementation of the World Summit on Sustainable Development*. United Nations, 72 pp.

United Nations (2006) *World Water Development Report 2: Water, a shared responsibility*. UNESCO, Paris, 601 pp.

United Nations Committee on Economic Social and Cultural Rights (2003) *General Comment No. 15 (2002). The Right to Water.* E/C.12/2002/11, United Nations Social and Economic Council, 18 pp.

UNDP [United Nations Development Programme] (2006) *MDG Targets and Indicators* [accessed 06.03.07: http://www.undp.org/mdg/goallist.shtml

World Water Council (2006) *Final Report of the 4th World Water Forum*. National Water Commission of Mexico, Mexico City, 262 pp.

Zobel, M. (1992) Plant-species coexistence: the role of historical, evolutionary and ecological factors. *Oikos*, **65**, 314 - 320.

Zobel, M., van der Maarel, E. & Dupré, C. (1998) Species pool: the concept, its determination and significance for community restoration. *Applied Vegetation Science*, **1**, 55 - 66.

Journal of Practical Ecology and Conservation Volume 7(2) 2009

The Assessment of Underground Water for Potable and Agricultural uses in the Elba Region of south-eastern Egypt

Mohammed Kamel Ahmed[1], Younis Abdoul Moula Alhendawi[2] and Ian D. Rotherham[3]

[1] Department of Botany, Faculty of Sciences, South Valley University, 83523 Qena, Egypt.
mohamed-kamel@lycos.com

[2] Faculty of Natural Resources, University of Omar Almkhanar, Tobruk, Libya.
alhendawiy@hotmail.com

[3] Sheffield Hallam University, UK
i.d.rotherham@shu.ac.uk

Abstract

During the period from 2002-2003, underground water samples were collected from ten locations, springs and wells, in the eastern desert of Upper Egypt the "Elba region" (Map 1). In the springs (such as in Abraq Spring 1, Abraq Spring 2 and Abo-Zafa), the water table lies at between 0.5–0.75 m below the ground surface. In the wells, the water table is deeper than the previous depth, for instance, the water table in Qumbeet well lies at 5 m. depth approximately. The water samples were analysed physically, chemically and biologically. The parameters measured during this investigation were temperature, pH, dissolved oxygen (DO), electrical conductivity, and the water contents of Na^+, K^+, Ca^{2+} and Cl^-. In addition to these the samples were assessed for selected microbiological parameters: Total Coliform (TC) and faecal Coliform (FC) densities, and Coli-phages "MPN" and Salmonella-phages. Additionally, moulds ("aquatic and filamentous fungi and yeast") and algae were determined in the water samples. The aim of the study was to evaluate the water quality and suitability for domestic home use and in agriculture. The results demonstrated that all the samples examined suggested that the water here is fit for both potable and agricultural uses.

Key words: Elba region, underground water, water quality, wells, springs

Introduction

Water is vital for life and development in all parts of the world. The need for water increases with population growth, so water will become a great problem especially in The Middle East and North Africa. In this case, the necessity to find other water sources additional to the Nile is an important issue. One of the most important resources across the region is the wells and springs distributed across the eastern desert of Egypt. These resources are noted as especially helpful for Bedouins, for domestic use, and for animals. Increasingly these wells and springs can be used for agricultural purposes and for increasing the extent of cultivated areas. The south-eastern corner of Egypt, especially the Elba region has several natural wells and springs together with a number of artificial wells. In considering the development of water resources in this region, the evaluation of water quality and suitability is important. One issue is to determine the degree to which these sources can be depended on for the purposes of agricultural or domestic usage. The groundwater resources originate mainly from occasional rainfall and associated torrents. This water partially infiltrates through the friable loose sediments to accumulate in basement depressions or be trapped by faults and buried dykes (Shalabi, 1987a and 1987b; Shalabi *et al.*, 1987). Chemical and biological characteristics of these waters are important in terms of their suitability for use.

Figure 1. Map of the Elba region indicating the locations of the studied wells and springs.

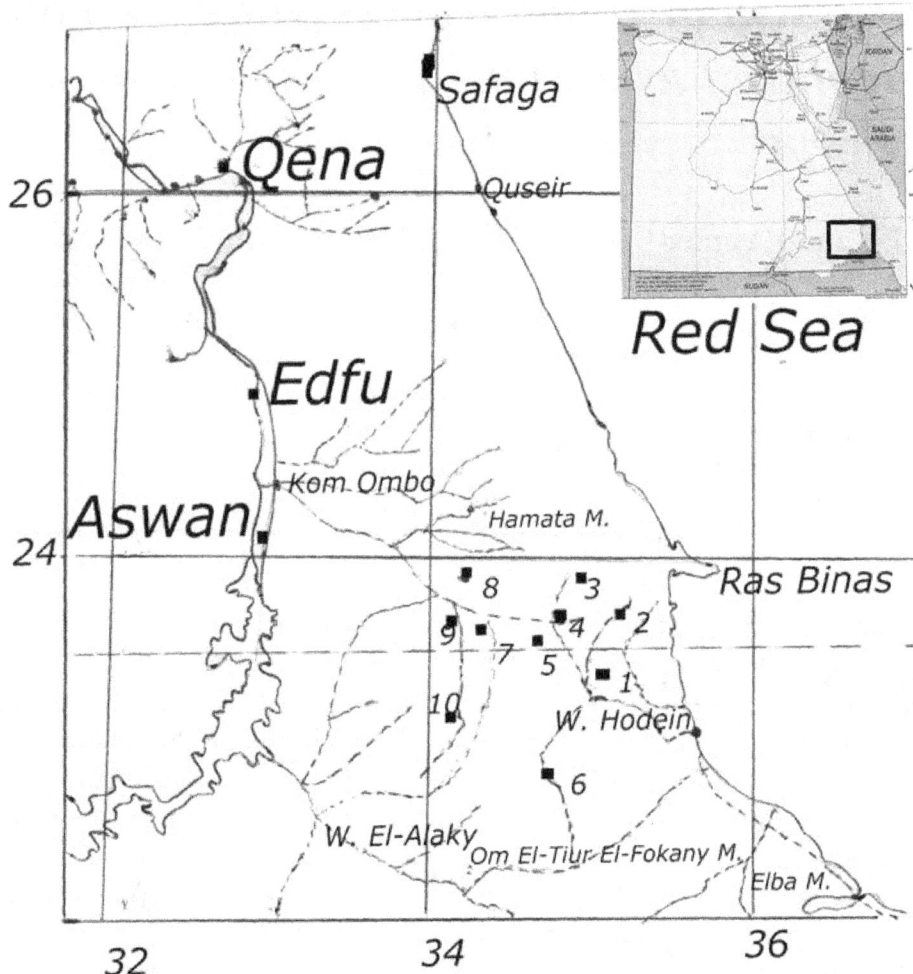

The most common method for bacteriological examination of water samples in relation to possible sewage contamination is through the use of the coliform group and / or faecal streptococci as indicators of pollution (Geldreich *et al.*, 1962; Mundt *et al.*, 1962; Geldreich *et al.*, 1971; El-Hawaary *et al.*, 1987 & 1992; Mohawed, 1994c & 1995). Whenever a groundwater resource is being exploited, the appropriate chemical and physical analyses of water samples should be considered as standard measures for suitability (APHA, 1985). Considerable attention has also been paid to fungal flora contained in different types of water resources in recent years (El-Hawaary *et al.*, 1992; Mohawed, 1993a & b), and it is considered important to assess this. Finally, the physical and chemical properties must be determined in order to evaluate the degree of salinity and toxicity in the waters.

The aim of this study was to undertake an evaluation of water quality in the well and springs distributed in the south of the Egyptian Eastern Desert, and in doing this to complete the survey begun by Mohawed in 1994. There are implications of these studies for human settlement, and for land use and agriculture.

Materials and Methods

The study area lies between latitudes 32° and 35° as shown in Figure 1 and includes two large wadis. The first wadi is called Wadi Hodein and is approximately 200 km long. The second wadi is called Wadi Qumbeet, and extends over 200 km. Temperatures range between 44°C in August and 26°C in January. The maximum average rainfall is about 1.4 mm in December, and the months from June to September are rainless.

Water samples were collected at three-month intervals during the period from April 2002 to April 2003. Samples for microbiological analysis placed in sterile glass bottles (500 ml), kept in an icebox during the journey, and then placed in a refrigerator at

-10°C in the laboratory. The samples for physical and chemical analysis were taken in 2-litre plastic bottles.

Some field measurements were also undertaken. These included temperature, pH (using a pH meter Model 5998-10, Horizon Ecology Co., USA), total dissolved solids "TDS" (using Nalcometer Model C.I.G.R., France), Dissolved Oxygen (using "Bibby" Dissolved Oxygen SMO1, Germany), electrical conductivity and salinity (YSI Model 33 S.C.T. meter, Yellow Spring Instrument Co. Inc., Ohio Model 33). Sodium, Potassium, Lithium and Calcium were determined in the laboratory by Flame photometry using a 410 "Corning M410 CIBA" (Corning Diagnostics Scientific Instrument, Halsted, Essex, England). Soluble chloride was estimated by applying the silver nitrate titration method using potassium chromate as an indicator (Jackson, 1985; Table 3).

Bacteriological analysis:

Total and faecal coliform densities by MPN test were carried out using MacConkey broth, for the presumptive test and the Eosin Methylene Blue agar "EMB" for confirmation (WHO, 1984, Mohawed, 1994).

Coliphages assay:

Coliphages were assessed using the procedure recommended by APHA (1985), Kott *et al.* (1974) and Ludoviciet *et al.* (1975).

Mould and yeast examination:

For detection of moulds and yeasts various media were used. For filamentous fungi the base was Dox–agar medium "Dox,1910" to which streptomycin "30 ug/ml" and aureomycin "25 ug/ml" were added to prevent the growth of bacteria. Gorodkowa medium was used for yeast assessment (Friedrich, 1962). The media were poured in a sterile Petri dishes and dried; then a 1 ml sample was inoculated onto the surface of each plate and incubated at 28°C ±2°C .The plates were examined within 48-72 hrs for yeast and five days for moulds.

Aquatic fungi:

For the isolation of zoosporic fungi associated with different water samples, seeds of hemp, maize, and beans were used. The sterile seeds were incubated in Petri dishes with water sample for seven days at 20°C at standard room daylight. Each colonised seed was examined, identified and transferred to another Petri dish with sterile seeds and under sterile conditions. Seeds as maize or beans were selected as a medium to germinate the zoosporic fungi. These seeds are the main crops that can be cultivated in the studied area.

Algal flora aggregate with water:

The composition of the phycoperiphyton (algae) communities of wells and springs which were present in this investigation were determined during the period from (4/10/2002 to 4/10/2003). The term periphyton is of Russian origin (Behning, 1928) and first referred only to organisms growing on objects placed in water by people. Epiphytic algae were collected and examined by compound light microscope.

The algal floras, taxonomy and nomenclature used were based on Pascher (1940) as employed by G. Smith (1944, 1950) and Behre (1956). The study of algal growth on the Nile system began seriously at Aswan and Sohage by El-Otify (1991) and Ali (1993).

Results and Discussion

Coliforms are several different types of bacteria that exist in the intestines of warm-blooded animals and are found in bodily waste, animal droppings, and naturally in soil. Coliform bacteria are described and grouped, based on their common origin or characteristics, as either total or faecal coliforms. The group of total coliforms includes faecal coliform bacteria, such as *Escherichia coli* (*E. coli*), as well as other types of coliform bacteria that can survive in soil and vegetation. Total coliforms do not necessarily indicate recent water contamination by faecal waste. However, the presence or absence of these bacteria in treated water is often used to determine whether water disinfection systems are functioning.

Table 1 shows the constituents of the media used for bacteriological analysis. This was conducted according to APHA (1985) using the five replicates fermentation tubes FT5 procedure.

Coliform bacteria are widely used internationally as indicators of pollution (Voelker *et al.*, 1960, Aboo *et al.*, 1986;

Mohawed, 1994 & 1995). They imply presence of other pathogenic micro-organisms, specifically organisms of Faecal origin. The total coliforms represent the whole group, and are bacteria that multiply at 37°C.

As shown in Table 2. the density of total viable bacterial counts ranged between 80-140 cells/ml and 110-150 cells /ml in incubation temperatures of 25°C and 37°C respectively. Borm (1974) stated that water is acceptable for animal use if the total bacterial count at 22 °C or 37°C is below 1,000 cells/ml. The

Table 1. The constituents of the media used for bacteriological analysis.

Number	Constituents g/L	Total Coliform	Faecal Coliform	Endo-Methy-Media
1	Lactose	10	10	10
2	Peptone	20	10	10
3	Sodium Chloride	5	-	-
4	Bile salt	-	20	-
5	Eusin	-	-	0.4
6	Brilliant Green	-	0.133X10.5	-
7	Phenol Red	+	-	-
8	Methylene Blue	-	-	0.065
9	K_2HPO_4	-	-	2
10	Agar-Agar	-	-	20

Enteric bacteria, total coliform (TC), and faecal coliforms (FC) in water samples from ten wells and springs were assessed by spread-plating 0.2ml on Eosin Methylene Blue Agar (Oxoid, 1982) and incubated in duplicate plates at 25°C and 37°C respectively.

Table 2. Density of bacteriological parameters of underground water in Elba regions

Name and Location	Type of Water Sources	Total Counts 25c	37c	Total Coliform MPN/100ml	Faecal Coliform MPN/100ml	Total Coliphage in 100ml
Abrak I	Spring 0.5-2m	12x10	13x10	4	0.0	0.0
Abrak II	Spring 1-2m	14x10	12x10	2	0.5	5.75
Kumdeep	Well 5-10m	12x10	14x10	6	0.33	5.75
Abo-Zafai	Spring 0.5-0.75m	13x10	14x10	9	0.75	5.75
Alaqraka-Eldiana	Spring	9x10	11x10	7	0.0	0.0
Alaqaka –Elastana	Well	12x10	13x10	17	1.5	11.5
El-Gahilia	Well	8x10	11x10	<2	0.0	0.0
Aukaba	Well 12-13m	11x10	12x10	9	1.0	0.0
Deef	Well	14x10	15x10	17	1.75	5.15
Akawa	Well	14x10	15x10	17	1.75	5.15

Table 3. The mean values of physichemical analysis from underground water in the Elba region.

Region / Name Parameter	Abraq 1 Spring	Abraq 2 Spring	Qumbeet Spring	Abo-Zafa Spring	Al-Erka El-Raiana	Al-Erka El-Atshana	El-Jahellia Well	Aukaba Well	Deef Well	Aqua Well
Temperature C	25	30	30	25	25	24	26	30	29	30
PH Values	8.04	8.09	8.04	8.01	7.70	7.70	7.25	8.01	7.09	7.66
Total salinites as (CaCo) ug/L	.567	.670	05.06	0.681	0.380	0.440	0.370	0.420	0.401	0.2020
Dissolved Oxygen (DO) ng/L	21.20	20.80	18.50	22.01	18.20	20.80	17.20	18.20	17.02	16.05
Electrical Conductivities (umbo/cn) or NaHos/cm	5.38 ms	6.11 ms	6.73 ms	5.50 ms	11.58 ms	12.75 ms	6.38 ms	11.83 us	30.78 ms	396 us
Na (ppm)	140	14.2	22.1	22.01	7.2	7.1	62.1	48	48.8	50
Ca (ppm)	150	160	154	150	170	160	68	146	140	220
Chloride mg/L	96	120	140	120	170	98	126	120	89	130
K (ppm)	1.7	1.6	3.3	1.4	1.2	0.7	0.8	0.8	0.8	1.2

Table 4. The number of different species and genera of fungi which were isolated from water samples collected from different springs and wells in the Elba region: (a) aquatic fungi.

Region / Name Aquatic Fungi Actasiomycetes	Abraq 1 Spring	Abraq 2 Spring	Qumbeet Spring	Abo-Zafa Spring	Al-Erka El-Raiana Spring	Al-Erka El-Atshana Spring	El-Jahellia Well	Aukaba Well	Deef Well	Aqua Well
Prosteliales	Protostelium mycophage	Protostelium mycophage		Protostelium mycophage	Protostelium mycophage	Protostelium mycophage				
Acrasiales										
Dictyosteliales	Dictyostelium discoideum	Dictyostelium discoideum	Dictyostelium discoideum	Dictyostelium discoideum	Dictyostelium discoideum	Dictyostelium discoideum				
not detected										

Table 4. The number of different species and genera of fungi which were isolated from water samples collected from different springs and wells in the Elba region: (b) other species and genera.

Water source	Aspergillus niger	Aspergillus terus	Aspergillus flavus	Aspergillus illidux	Aspergillus sydaws	Aspergillus	Total Count	Penicillium notatum	Penicillium chrosa	Penicillium citri.	Penicillium oxal.	Total Count	Rhizopus sp.	Paciolomyces sp.	Fusarium sp.	Syncephalos-trum	Alter sp.	Total Count	Candida albicans	Candida solani	Cryptococcus	Cryptococcus	Cryptococcus ned.	Rod. raba	Rod. Fla.	Soc. ar.	Tor. fam	Tor. candia	Total Count
Abraq Spring I	2	2	1	1	1	2	9	1	0	1	1	0	1	0	0	2	3	14	3	2	3	2	4	5	1	2	1	1	9
Abraq Spring II	2	4	1	2	2	3	14	1	0	2	1	0	1	0	1	3	2	13	2	3	2	3	3	6	2	1	1	2	11
Qumbeet Spring	0	0	1	3	2	1	7	1	2	1	2	0	1	0	1	2	3	14	3	2	2	4	3	7	1	2	1	3	13
Abo-Zafa Spring	1	1	1	2	3	4	12	4	2	1	1	8	4	3	3	1	2	13	4	1	1	1	0	0	0	1	0	1	9
Al-Erka El-Raiana Spring	4	3	2	0	4	2	16	3	2	2	2	9	4	2	2	2	3	13	2	2	1	0	1	0	1	0	1	1	9
Al-Erka El-Atshana Spring	3	2	1	2	1	2	9	4	2	1	2	9	2	1	1	1	2	7	0	1	1	2	1	0	2	0	0	1	8
El-Jahelia Well	0	0	1	0	0	0	1	2	4	3	1	10	1	1	2	0	0	4	1	2	1	2	1	1	0	1	1	0	10
Aukaba well	2	1	2	3	4	5	17	3	2	3	2	10	0	1	0	1	1	3	2	1	0	1	2	1	2	1	1	0	11
Deef well	2	6	5	4	5	4	26	3	4	2	1	10	1	1	0	0	0	2	1	2	2	1	0	0	1	1	2	1	11
Aqua well	1	1	2	3	5	2	14	4	2	2	1	9	2	1	1	1	1	6	2	2	1	1	0	1	2	2	1	1	13
Not detected	25	27	19	17	20	17	125	29	21	19	14	83	24	19	16	13	17	89	20	18	9	8	8	3	12	10	8	8	104

Genera and Species

WHO (1984) levels acceptable for drinking water are less than 300 cells/ml. The total coliform ranged between 2 and 17 (MPN/100ml) while Faecal coliform did not exceed 1.75. The Faecal coliform group of bacteria has been used as an indicator of water quality with respect to the presence of human pathogens.

Coliphages have been proposed as microbial indicators as they behave more like the human enteric viruses that pose a health risk to water consumers if water has been contaminated with human faeces. As shown in our data, coliphages ranged between zero in two wells and two springs, as for instance Abrak I Spring, and reached up to 11.5 in Alaqaka–Elastana well. According to the standard methods and criteria (APHA, 1985), the samples were suitable for drinking and for public water supplies. Coliphages correlated positive significantly at 0.05 versus total coliform, at 0.01 versus Faecal coliform. This is agreed with Guelin (1948 and 1950).

Assessments of the physical characteristics for the tested water resources are presented in Table 3. These include the pH range from 7.09–8.09, and water temperature from 24-30°C. Dissolved Oxygen (DO) from the sampled sites always showed relatively high values because most of the data were collected near the ground surface. This was also influenced by the algal flora, so, DO ranged between 16.05 and 22.01 mg/L. The electrical conductivity and estimated ions (Table 3) value laid under the permitted international levels.

It was observed that various fungi were present and isolated from all samples (Table 4). *Aspergillus* was the most common genus with *Aspergillus* group being isolated in 125 samples. Yeast flora scored 104 in isolates from 0-20 per/ml sample (Table 4), the most common being *Candida albicans* which scored 20. It should be noted that slime moulds and similar organisms are sometimes considered as "Aquatic Fungi". This view is shared by Olive, (1970) who reviewed the classification of the groups. *Achrasiomycetes* have only recently been discovered probably they are inconspicuous. However, they are ubiquitous on decaying plant materials, in soil, and in freshwater. Olive (1967) provided a monograph on this group. They can be cultivated on weakly nutrient agars or by hay infusion agar. They can be ingested as food in conjunction with other organisms such as bacteria *Escherichia coli, Klebsiella aerogenes*, and the yeast *Rhodotorula mucilaginosa*. With this is mind the *Prosteliales* have been suggested as an indicator of water pollution and as possible models for assessing enterovirus removal and inactivation during treatment of drinking water and waste water.

The results indicated that the phycoperiphyton populations and algal flora in many springs belonged mainly to four groups: *Chlorophyta, Cyanophyta, Bacillariophyta* and *Pyrrhophyta (Euglenophyta)*.

Finally, in summation, our data showed that there was a correlation between both total coliform and Faecal coliform versus total coliphages (Guelin, 1948, 1950). The bacterial count was under the acceptable level recommended by WHO (1984). The estimated physical characteristics as mentioned above were in the international safe range. The presence of aquatic fungi and algae in the studied water samples can play an important role in purification processes (Batko, 1975; Muller & Loeffler, 1987). Physical and chemical processes of water self-purification are often regulated by biotic factors or strongly depend on them (Ostroumov, 2001, 2002). So, the included analysis for the different studied samples showed that the water resources can be used safely, but the continued assessment is necessary to avoid any unexpected pollution in the future.

Conclusions

This study presents an assessment of water quality in selected wells and springs in south-eastern Egypt. These sites represent the main sources of water for local Bedouins. The analysis of data from all the tested springs and wells indicates that the water resources are safe for the various uses that are desired. According to established benchmarks such as WHO guidelines the waters from all the sites tested are safe to be used for drinking and other domestic purposes. Furthermore, in this arid landscape, these water resources can be used for agricultural irrigation. This paper helps to develop the emerging literature on this important topic.

Acknowledgements

The authors would like to thank all the staff members of the protected area of Elba for their kind attention and helpful assistance during the study.

References

Aboo, K.M., Sastry, C.A. and Alex, P.G. (1986) A study of well waters in Bhopal City. *Environmental Health*, **10**, 189–2003.

Ali, Z.A.M (1993) *Ecological studies on the Algal flora of Egyptian Soils Sohage district.* Unpublished MSc thesis, Faculty of Sciences (Sohage), Assiut University, Egypt.

APHA (1985) *Standard Method for the Examination of water and wastewater.* 16th American Public Health Association, Washington DC.

Batko, A. (1975) *Hydromycology – an overview.* Państwowe Wydawnictwo Naukowe, Warszawa. (In Polish).

Behning, A. (1928) *Das Lebender Wolga. Zugleich eine Einfuhrung in de Fluss-Biologie.* In: Thienemann, A. (Ed.), Die Binnengewasser V., Stuttgart, 162 pp.

Behre, K. (1956) Die Algenbesidlung einiger seen um Bremen und Bremerhaven. *Ver. Inst. Meersk. Bermmerhaven*, **4**, 221-283.

Behre, K. (1961) Die Algenbesiedlung der Unterweser unter Berücksichtigung ihrer Zuflüsse. *Veröff. Inst. Meeresforsch., Bremerhaven*, **7**, 71–263.

Borm, F. (1974) *Chemical and Bacteriological Examination of Drinking Water.* Committee on Water Quality Criteria (1972), National Academy of Sciences, Washington DC.

Dox, A.W. (1910) The intracellular of Penicillium and Aspergillus with special reference to those of P.camenberti. *U.S. Dept. Agr. Animal. Ind. Bull.*, **120**, 170.

El-Hawaary, S., Sokker, I.M., Ali, M.A. and Khalafalh, G.M. (1987) Viruses in treatment effluents 1- Drinking water. *Journal of the Egyptian public Health Association, LXII* ; **5**, 6.

El-Hawaary, S., Kamel, M.M., Ali, M.A. and Mohamed, G.E. (1992) *Assessment of water quality in rural areas (ATIRICA-GIZA.* Proceedings of The First Middle East Conference on Water Supply and Sanitation for Rural Areas, February 23-25, 1992, 145-152.

El-Otify, A.M.A.H. (1991) *Studies on phycoperiphyton of the Nile system at. Aswan, Egypt.* Unpublished PhD thesis, Botany Department, Faculty of Science, Aswan–Assiut University, Egypt.

Fridrich, E. (1962) *Sprosspilze des Menschen. Johann Ambresius*, Barth, Veriag, Leipzig.

Geldreich, E.E., Bordner, R.H., Kuff, C.B., Clark, H.F. and Kabler, P.W. (1962) Type distribution of coliform bacteria in the faeces of warm blooded animals. *Journal of Water Pollution and Control Federation*, **34**, 195.

Geidreich, E.E. (1971) General Microbiology. *Journal of Water Pollution and Control Federation*, **43**, 1202.

Guelin, A. (1948) Etude des bacteriophages typhiques. Vidans les eaux. *Ann. Inst. Pasteur Paris* **75**:485-496.

Guelin, A. (1950) Sur le choix des souches etalons pour la detection du bacille typhique dans les eaux par la recherché des bacteriophages specifiques. *Ann. Inst. Pasteur Paris* **79**:186-191.

Hoffman, L. (1989) Algae of terrestrial habitats. *Botanical Review*, **55**, 7–105.

Jackson, M.L. (1985) *Soil chemical analysis.* Constable and Coilondon Ltd., London.

Kott, Y.N., Roze, S.S. and Betzer, N. (1974) Bacteriophages as viral pollution indicators. *Water Research*, **8**, 165.

Ludovid, P.P., Phillips, R.A. and Jeter, W.S. (1975) *Comparative inactivation of bacteria and viruses in tertiary-treated wastewater, by* chlorination. In: Johnson, J.D. (ed.) *Disinfection water and wastewater*. Ann. Arbr. Science. Publ., ANN. Arbor. Mich.

Mohawed, S.M. (1994a) *Isolation and identification of some Iron bacteria from sedimentation sludge on the River Nile in Upper Egypt, Islands*. Proceedings of National Conference on the River Nile, Assiut University Center for Environmental studies (AUCES), pp 75-88.

Mohawed, S.M. (1994b) *Studies on the Algal flora and physico-chemical character on sedimentation sludge on the River Nile*. Proceedings of National Conference on the River Nile (AUCES), pp 61- 73.

Mohawed, S.M. (1994c). Assessment of Microbiological quality of underground water of Upper Egypt, Qena region. *Egyptian Journal of Microbiology*, **29** (2), 183- 192.

Mohawed, S.M. (1995) Assessment of underground water for potable uses in Eastern Desert of upper Egypt. *Egyptian Journal of Microbiology*, **30** (1), 161- 175.

Muller, E. and Loeffler, W. (1987) *Mycology – an overview*. Państwowe Wydawnictwo Rolnicze i Leśne, Warszawa. (In Polish).

Mundt, J.O., Cogyin, J.H. and Johnson, L.F. (1962) Growth of Streptococcus faecalis var. liquefacient on plants. *Applied Microbiology*, **10**, 552-555.

Olive, L.S. (1967) The protostelida: a new order of the mycetozoa. *Mycologia*, **59**, 1-29.

Olive, L.S. (1970) The mycetozoa: a revised classification. *Botanical Review*, **36** (1), 59-89.

Ostroumov, S.A. (2002) *Biologicheskie effekty pri vozdeistvii poverkhnostno-aktivnykh veshchestv na organizmy (Biological Effects of Surfactants)*, Moscow: MAKS-Press.

Ostroumov, S.A. (2002) On the Role of Hydrobionts in Regulating the Fluxes of Matter and Migration of Elements in Aquatic Ecosystems, *Vestn. Ross. Akad. Est. Nauk*, **2** (3), 50–54.

Oxoid Ltd (1982) *The Oxoid Manual Book*. Fifth Edition, Basingstoke, Hampshire, UK page 133.

Pascher, A. (1940) Zur Kenntnis der SiJsswassertetrasporalen I. Beih. *Bot. Central. Abt.*, **60**, 135-156.

Shalabi, M.S. (1987a) Evaluation of the groundwater aquifer at the entrance of Qift –El-Qoussier road Eastern Desert of Egypt. *Bulletin Faculty of Sciences, Qena, Egypt*, **1** (1), 173.

Shalabi, M.S. (1986b) Evaluation of the setting of aquifer, on Qena–Safaga road, Eastern Desert of Egypt. *Bulletin Faculty of Sciences, Qena, Egypt*, **1** (1), 185.

Shalabi, M.S., EL-Hosseini, A.H. and EL-Khoteeb, S.O. (1987) Groundwater possibilities in Wadi Qena, Eastern Desert of Egypt. *Bulletin Faculty of Sciences, Qena, Egypt*, **1** (1), 143.

Smith, G.M. (1950) *The freshwater algae of the United States*. 2nd. Ed., MacGraw–Hill, New York p719.

Smith, G.M. (1944) A comparative study of the species of Volvox. *Transactions of the American Microscopical Society*, **63**, 265 –310.

Voelker, R.A. and Heukelekian, H. (1960) Seasonal coliform variation in waters. *American Journal of Public Health*, **50** (12), 1873–1881.

WHO (1984) *Guidelines for Drinking Water Quality. Vol. 1 Recommendations*. World Health, Organization, Geneva.

Visualizing impacts of land use change on flooding

Fergus Sinclair[1], Karen Padmore[2], Thomas Varsimides[3], Tim Pagella[1], Bethana Jackson[4], Amy Eycott[5] and Barbara Orellana[4]

[1] School of the Environment and Natural Resources, Bangor University
[2] Technium CAST, Bangor
[3] School of Computer Science, Bangor University
[4] Department of Civil and Environmental Engineering, Imperial College London
[5] Ecology Division, Forest Research, Alice Holt

Notes

Introduction

Human behaviour that influences flood risk management typically involves a range of different stakeholders in both source and sink areas. For most of these people, flooding is either an infrequent event, or a distant, downstream consequence that is difficult to associate with their actions. Here, we explore the use of visualisation techniques to engage different stakeholders in thinking about how changes in land use may influence flood generation. We demonstrate three approaches that have been used to develop practical tools aimed at different audiences and outcomes.

Animation of general principles in flood generation

Urban flood risk may be impacted by land cover within the urban area itself, in terms of making space for water, as well as by land use in the surrounding countryside. When discussing urban development or potential land use change upstream with key stakeholders, including urban residents, commercial interests involved in development, and land owners in the countryside, a key issue is the extent to which people have a similar understanding of flood generation processes. Here we use dynamic visualisations to illustrate basic principles of how changes in land cover can influence flooding. These can be used as a basis for grabbing and holding the attention of stakeholders and for societal learning about these processes. They are effective tools for use as a precursor to planning negotiations.

Landscape visualisation

Land use in the Welsh uplands may influence river flows locally within Wales and along transboundary rivers that flow into England, like the Severn, but farmers and other land owners rarely consider the effects of their decisions about land use on catchment hydrology. There are three major reasons for this:

- water flow and quality at a local scale has often not been of much interest to land users nor has it generated value for them (this may change as the EU Water Framework Directive is implemented);

- decisions about land use are made at field and farm level but stream and river flows are manifest at a catchment scale (so that unless social capital exists at a catchment scale amongst farmers, management of water courses may not be possible);

- decisions about land use change that affect hydrology are made for multiple reasons, most of which are not associated with flood risk management (they are also affected by a range of drivers, including policy instruments, that are rarely integrated at the landscape scale at which key ecosystem services such as river flows are manifest).

These issues are generic problems that need to be overcome for impacts of upstream land use on downstream flooding to be considered at source. As part of the Flood Risk Management Research Consortium

(FRMRC) we have developed and piloted a landscape visualisation tool for the Pontbren sub-catchment in mid-Wales. This area was chosen because, uniquely, a group of ten farmers with contiguous land have got together to manage some aspects of their farming in collective ways that make it possible to consider sub-catchment management.

Jimile, the landscape visualisation tool that we have developed, is designed to be used for three separate tasks:

- for policy makers and land users to learn about effects of changing policy and/or land use on catchment hydrology;

- for participatory planning of land use policy at a landscape scale; and

- for spatially explicit negotiation of specific land use change.

The tool comprises a landscape visualisation that can be connected to models of effects of land use change on catchment hydrology. Animation of the landscape (such as river flow) can either be driven by a simple realtime simulation or from data streams generated by sophisticated hydrological models run for predetermined alternative scenarios. Different land use scenarios (such as different amounts and spatial configuration of tree cover) can be dynamically created and modified by the user, and the land use change and its impacts on hydrology can be viewed in either a two or three dimensional rendition of the landscape.

The tool has been immediately recognised as a powerful way of engaging stakeholders and enabling spatially explicit consideration of how land use change affects hydrology.

Generalisation

Key developments identified as desirable from using *Jimile* have been:

- an ability to trade off impacts of land use change on hydrology, with effects on other ecosystem services such as biodiversity and productivity; and

- to be able to set up and run visualisations for any catchment, using readily available data.

We have begun to address these requirements with funding from the EU Robinwood interregional project, by developing a tool for planning and negotiating tree cover at landscape scale that overlays the output from a simple flow accumulation model based on topography and land use, with habitat networks and farmers assessments of where tree cover may enhance or conflict with productive use of land.

Book Reviews

The Naturalized Animals of Britan and Ireland.

Christopher Lever

2009, New Holland Publishers (UK) Ltd, London, 424 pages

Hbk. £35 ISBN 978-1-84773-454-9

For those undertaking habitat conservation management some background knowledge of the impacts of exotic (alien) animals is very helpful. To have this all within one cover is a great bonus. So here we have a timely and much needed update and reissue of a classic book first published in the 1970s. Lever presents a thoroughly readable and authoritative account which details how many alien vertebrate animals (mammals, birds, reptiles, amphibians and fish) were introduced to great Britain, became naturalised and are now living wild. He discusses species released deliberately as well as those which escaped accidentally. Based on many years of painstaking research the book describes the individual histories and also their impacts on the environments into which they have become embedded; fascinating stuff.

The story, or perhaps stories, of animals introduced to Britain covers many centuries but the most abundant examples are from the nineteenth and twentieth centuries as landowners and others sought to diversify the native fauna. British winters proved a challenge to many species and a lot did not survive. The Peak District wallabies for example, were lost because of a combination of recreational disturbance and then the series of cold 1980s winters. However, many species did establish, and quite a few, such as for example little owl and American mink, thrived in their new found homes. Some of the newcomers were eventually deemed beneficial; so after warnings of dire consequences for game birds, little owls are now considered by farmers to eat insect pests and so are welcome additions to our ecology.

Others like the mink have devastated populations of native water voles and other mammals and birds.

There is a whole diversity of animals introduced at different times and for a variety of reasons or motivations. From common carp to red-necked wallaby and from ring-necked parakeet to African clawed toads British ecology has acquired and absorbed and them all. At the same time, as British fauna has gained exotics it has also lost many natives, mostly through habitat loss and persecution. Some species were lost very early, but their demise has affected 'native' ecology ever since. They include all our larger carnivores, plus beaver, wild boar, and birds such as white-tailed eagle. The red kite was largely extinct in most of its British range but a reintroduction programme has brought it back from the brink. Lever has extended the 1970s account to include the reintroductions, and of course these, like the naturalisations, are potentially controversial. It has been advocated that keystone carnivores such as lynx and wolf should be brought back, but many of the public and often landowners are filled with trepidation. Even the humble beaver raises the levels of concerns amongst river owners. This new edition of a unique book addresses all these matters with authority combined with a lively written style. The accounts are well researched and clearly written to provide an excellent case-book for these species, their histories and their impacts. For every taxonomic group Lever provides a rich source of research, of literature and of history to illuminate the story. The book takes the fauna, species by species, and gives detailed accounts with useful summaries of information on, for example, distributions, and of key literature. The national distribution maps are also clearly presented and very helpful.

The book gives accounts of the mechanisms and needs for potential control of some species, for example Canada goose, and of course the ruddy duck. For the latter species, Lever gives a detailed and hugely informative overview. The book also has a series of useful appendices on dates of earliest arrivals, on distributions and other background. There is a very helpful and comprehensive reference list too.

The first edition of this book was hugely important and I think notable for the breadth of knowledge and research as well as for the clarity of writing. Lever was also just about the first person to bring to a wider audience the impacts of the acclimatization societies, and for that alone deserved to be read. However, the new version does more that simply bring the book up to date, it moves it on considerably in terms of both information and presentation. The current edition is very nicely presented and at £35 for the hardback, at present-day prices, is excellent value. A question with a re-edition of a book, even a classic, will always be whether the new edition justifies its purchase. Well in this case the answer is a massive unequivocal yes! If you didn't read the original then you've certainly missed out, but either way you should rush out and buy this one. Christopher Lever has written a real treasure trove and you will read this again and again. It is a text that you will pore over once, but you will also find yourself returning to in order to dip into the little gems of the species accounts. What we need now is the same for British exotic flora; now there's a challenge.

Ian Rotherham

Nature Conservation Law.

Colin T. Reid. 2009, W. Green, Edinburgh, 435 pages

Hbk. £55 978-0-414-016958

Wildlife and especially species protection law is particularly pertinent to ecological consultants and professionals, and this extends beyond the obvious such as the protection for old trees or for roosting bats etc. It is worth considering that if your work impacts on a Badger sett for example, or you are clearing riverside trees for a local authority or the Environment Agency perhaps and affect a Kingfisher nest site or a Water Vole habitat, then the onus is on you to be within the law. Ignorance is no excuse and the cost of failure to comply can be very high. Practitioners often assume that the agency or other body that has commissioned their work will have checked these things out, but I would advocate caution before accepting this. If it hasn't been done, then it is you or your colleague that will be in breach. Anyway this is a short introduction as to why you might wish to read or refer to this book.

This is the third edition of a very important text by Colin Reid who is Professor of Environmental Law at the University of Dundee. Unfortunately with legal matters the situation is constantly changing and so there is a need to keep your literature and guidance up-to-date, and by the very nature of the need for a substantial comprehensive text, this is not cheap. First published in 1994 this third edition updates the second in 2002. However, despite the price this book will be a worthwhile investment for any practitioner whose work may stray into areas of the need to comply with environmental or nature conservation law.

As the preface explains this is not a book about nature conservation policy and practice but about the associated legal restrictions and interventions. It deals with the legal rules that penalise certain conduct and which also establish public authorities and gives them powers to intervene in the ways in which individuals and organisations treat wildlife and the land and waters that support it. This new edition deals with the specifics of new legislation which covers just Scotland and the implementation of European directives. The issues of devolution and the additional complexities they bring are an important contribution of this book to the practitioner library.

In brief the book provides an overview of the history and background to current legislation and legislative frameworks and introduces a wide range of other context issues. It then introduces relevant organisations and structures dealing with nature conservation and leads into the protection of wild animals, the exploitation and destruction of wildlife, the conservation of habitats, and the law relating to plants. The final chapters deal with European and other international aspects of wildlife law, and then other miscellaneous matters. The book finishes with a series of very helpful and informative technical appendices. Sometimes the sheer amount of information can make using the book difficult and the indexing is not always very intuitive. However, if the book aims at the non-specialist practitioner then it would be good to have a cross-referenced section between for example species and their protection with a summary, and with the requisite sections in the book. I tried this with Great Crested Newt for example and it wasn't easy to locate the guidance. However, I did find out, and I believe usefully so, that the Scottish Amendment to the Conservation Regulations of 1994, and which implement the European Habitats and Species Directive, make it an offence to deliberately or recklessly harass an individual or a group of Great Crested Newts. So next time you are about to harass a newt in Scotland then think again.

Ian Rotherham

Trees for All Seasons: Broadleaved evergreens for temperate climates.

Sean Hogan

2008, Timber Press, London, 336 pages

Hbk. £25 9780881926743

The book is well produced and nicely illustrated with 370 full colour plates. The author is an experienced plantsman based in North America and has set out to produce what the publisher claims to be *'the ultimate guide to trees that provide four-season beauty'*. There are detailed descriptions of over 300 choice trees to use with guidance on suggested selections for a wide range of climatic situation. The book is billed as the first and only volume to be dedicated to broadleaved evergreens and I have to take their word for that; I really don't know if it is!

Hogan has chosen trees from all around the world with the intention of providing beauty and ornament throughout the year in a range of climate conditions. These broadleaved evergreens are supremely versatile and attractive garden plants and the guidance should enthuse and inspire gardeners and the enthusiasts to experiment more. The book provides information on drought tolerant species and cold tolerance ones too, and so this will help the planter to use the species most suited to their own circumstances. So even if its only Hollies and evergreen Oaks, there's something here for every one.

Ian Rotherham

Protecting Our Orchard Heritage: A good practice guide for managing orchard projects

Written by Ida Fabriozio and edited by Kath Dalmeny and Jeanette Long-field

Published 2008 by Sustain, London. ISBN 978-1-903060-46-9 pbk

N.B. All printed Sustain publications are £10 (plus post and packing: £1 - UK, £3.50 - Europe/USA, £5 - World).
For e-publications they suggest a donation of £10, which will allow Sustain to continue to produce such reports.

Orchards were formerly a significant part of the British landscape with major apple, pear and plum growing regions. Indeed, almost every farm, country and suburban house had its own fruit trees and they were impoertant to local peole and to local economies. Today apples are far and away the most popular British fruit but 60 per cent of our apple orchards have been destroyed since 1970. Losses of traditional orchards are actually because official statistics are based amlost entirley on commercial orchards alone. In recent decades EU grants have even been paid to grub up now 'derelict' orchards, with two thirds of the apple orchards lost in under 30 years. The main causes have been urban development, the need for more profitable crops, and cheap imported fruit from overseas. These have caused the loss of many small orchards. Kent, known as the "Garden of England", lost 85% of its orchards in the last 50 years. The ancient Herefordshire orchards has only 10% left. Devon has faired even worse with 90% lost since the 1940s. Loss of orchards has been in parallel with increased intensive production in those remaining. This is at the expense of wildlife and of different traditional fruit varieties. Bear in miond that there are up to 6,000 varieties of dessert and cooking apples and hundreds more cider apples but today only ten varieties account for 92% of the British eating-apple orchards. The two dominant British are varie-

ties the Cox and Bramley though at the peak of British apple season, around a dozen apple varieties are sold in larger supermarkets. Many stores stock more varieties 'out of season' with over two thirds imported from countries such as Spain, France, USA, China, South Africa and New Zealand. These varieties include Pink Lady, Royal Gala, Golden Delicious, and Granny Smith. In supermarkets it is rare to see traditional varieties like Discovery, D'Arcy Spice, Bess Pool, Peasgood Nonsuch, Marriage maker, Lord Lambourne, Bleinheim Orange, Dr Harvey, or Tydeman's Late Orange. These are worth stocking if only for the names! Some of these names derive from local histories of apple trees, communities and cultures from an age before the globalisation of the food economy. Many of the varieties had specific local, culinary and seasonal uses. Alongside the aples are hundreds of varieties of damsons, plums, cob nuts, cherries and pears. Most are rarely seen today and the race is on to safeguard this heritage for the future and its past..

During 2006 and 2007, Sustain worked on a national orchard project with *Leader+* (European Union funded programme of rural development) to conserve and bring into sustainable management traditional orchards in six EU *Leader+* funded areas: Hereford Rivers; Somerset Levels & Moors; Teinbridge; North West Devon; Mid Kent, and Cumbria Fells & Dales. At the local level this funded and supported activities based on locally agreed priorities to keep traditional orchards as part of the local landscape and economy and enjoyed by local people. At the national level, Sustain facilitated a network of local projects to promote mutual support and learning through an email group, website, annual events and dissemination of successful case studies. They also supported local co-ordinators in efforts to include orchards in policies for landscape, agriculture, planning, nature conservation and recreation. The project culminated in the production of an inspiring good practice guide which includes case studies of *Leader+* funded orchard projects and examples of other projects from around the Britian and continental Europe.

The guide is called *Protecting our Orchard Heritage: A good practice guide for managing orchard projects* and it is packed full of ideas, information, photographs and contacts. The purpose of the report is to help current and future orchard projects, supporting practitioners and local communities to enhance, protect and celebrate orchards. It gives practical advice on setting up and running orchard projects; looks at creative ways to engage local communities and the media; and explores how to make orchards more financially viable through fundraising and by selling orchard products. It draws upon many successful examples of Leader+ and similar projects and the advice and experience from orchard groups around the UK and elsewhere. This guide aims to inspire orchard enthusiasts everywhere to take practical action to conserve our orchard heritage for the enjoyment of generations to come.

Dawn Turner Programme Manager at Herefordshire *Leader+* states that *"The finished (report) is absolutely fantastic, a great balance of enough information to make it really useful but not so much as to make it cumbersome or hard to navigate. We've had really good feedback from people who have seen it so far"*

This is a great little guide and is available either as paper copy or to download. For a donation of £10 plus postage it really is great value. I've included the detailed contents below to give an idea of the scope. This is aimed at managers and volunteers in orchard projects, but will benefit all those interested in tree conservation and especially arboriculturists called in the deal with fruit trees and their problems.

Chapter 1: Introduction
Traditional orchards are under threat
How can orchard projects help reverse this trend?
Background to the Orchard Co-operation Project

Chapter 2: Finding out more about your orchard
i) Orchard history, heritage and culture

The importance of history and heritage
Orchard heritage and how to research it
Orchards and archaeology
ii) Orchards, wildlife and crop diversity
Wildlife
Crop diversity
iii) The landscape
The significance of orchards
Protecting orchards from development
iv) Orchard mapping
Why is mapping important?
Useful mapping advice
v) The financial value of orchard products

Chapter 3: Exploiting the many benefits of orchards
i) Educational activities
Running courses to promote orchard skills
Taking training a stage further
Working with schools
Integrating orchards into the school curriculum
Orchards and healthy eating in schools
Raising awareness through arts and theatre
ii) Community well-being
Community orchards
Health and the wider community
Community food projects
Amenity and public access
iii) Restoring and safeguarding biodiversity and landscape
Promoting genetic diversity
Official designations that aid conservation
Orchard management guidance
iv) Creating a thriving local economy
Adding value
Buying fruit processing equipment
Foods and branding linked to 'landscape character'
'Protected Name' schemes
Exploring different outlets for produce
More ways to gain support for local orchard produce
More than fruit and nuts
Attracting orchard tourists
Hiring out orchard space to generate income
'Sponsor a Tree' schemes and crop sharing
What type of business?

Chapter 4: Attracting funding and other support
i) Issues to consider and sources of funding

Funds from the public
European, national, regional and local support
Charitable trusts and foundations
The National Lottery
Support from local businesses
Measuring success
ii) The importance of good communications
Events and festivals
Publicity
Engaging the community
Interpretation for visitors
Generating media support
Starting a campaign

Chapter 5: Looking to the Future...

Appendix 1: Business plans and funding applications
Appendix 2: Traditional Orchards in the UK Biodiversity Action Plan
Appendix 3: Contacts and useful information

There is lots of information on contacts and sources of help or guidance and Sustain also have their own very helpful website: http://www.sustainweb.org

Ian Rotherham

Between Earth and Sky:our intimate connections to trees.

Nalini M. Nadkarni

Published 2008 by University of California Press, Berkeley, Los Angeles and London. ISBN 978-0-520-24856-4 Price £17-95 hbk

This is a lovely book and a great read. It is both well written and tightly presented. The author is a 'canopy biologist' which means she spends a lot of time climbing and studying trees, especially in the tropics but generally around the world. She mixes science, literature and poetry to create a richly described and illustrated account of interrelationships with and around trees.

The story covers more mundane topics such as 'What is a Tree?' but moves swiftly through the goods and services provided by trees, the heath–related aspects, and then onto things such as 'Signs and Symbols', and 'Spirituality and Religion'. There's much more but I recommend that you read it yourself! Perhaps for those tree professionals who after a long day in inclement conditions with a problem tree and a difficult client have retired to the local hostelry to contemplate why on earth they are doing this, here is perhaps the answer.

Ian Rotherham

Guidelines for Contributors

Submitting a Proposal

Please submit your proposal for a paper, note or book review by sending your details, a title and short abstract to info@hallamec.plus.com in the first instance. We will then send you more detailed instructions and a pro-forma for you to use. General guidelines are set out below for your information.

General Guidelines

Article Title

Should be informative and as succinct as possible and should reflect the content and ecological significance of the paper.

Abstract/Summary

Should state clearly and briefly the object of your study, the methods used, the results obtained and your conclusions. It should not exceed 5% of the length of your paper, nor should it introduce ideas or information not in the text.

Introduction

This sets the scene for your paper and should say why you did the work. If the work follows from previously published papers, a brief statement with a few key references will suffice. If the paper introduces a new topic, a more detailed argument may be needed.

Materials and Methods

The purpose is to explain clearly how you carried out your study. Make sure that there is a logical flow to the explanation, using sub-headings if necessary.

Results/Conclusion

The text of this section should focus the reader's attention on the features that you regard as being most important. You should present your findings in a logical order.

Discussion

The objective is to place your findings in the context of previous studies and the present state of the subject. Expressing opinions on the value, validity and veracity of both your work and that of others should be backed by evidence. Make it as precise and concise as possible.

Acknowledgements

Optional section. If included, please keep as brief as possible.

References

References cited in the text should be written in the following format: White (1972); White (1972a, b); White and Black (1972); White, Black and Grey (1972); (White, 1972); (White, 1972, 1973); (White, 1972; Black, 1973). For three authors all names should be given; for more than three use *et al.*

List references at the end of the paper in alphabetical last name order, giving the journal titles in full. Ensure that all references are cited correctly as they will not be checked by the editors.

A paper may only be cited as 'in press' if it has been accepted by a journal, and the journal should be cited in the References. Papers not yet accepted should be cited in the main text as 'unpublished', along with the initials of the author(s) (not as 'in preparation'), and omitted from the References.

Submission Instructions

1. Papers should be submitted as electronic copy using the pro-forma which will be sent to you. This should preferably be by email but if there are a large number of photographs, diagrams or illustrations we will ask for the full paper and accompanying photographs or illustrations to be submitted on a cd-rom with a text only version sent by email. The copy should be in .doc (Word) format. If a Macintosh computer is used, please ensure that the files can be read by a Windows PC. If using a version of Word that saves documents as .docx, please save this as a .doc file.

2. Set the paper size to A4 portrait, unless tables are large when landscape orientation can be used. Leave a margin of 2.5cm on all sides of the document. All the pages of the article must be consecutively numbered in the middle of the page. DO NOT use any other header or footers within the document.

3. At the start of the document, give the full article title, the name(s) of the author(s), and the address(es) of the author(s) at the time of writing the article. The present address (if different) should be included. In the case of multiple authors, please indicate the name and address of the author who will receive any comments and who is to check the proofs.

4. Keep the text in a basic text-formatting style only, using bold and italics where required. Use only UK English spelling, not US English, except for direct quotes. For tabulated data, please use the table function in Word or an Excel table, DO NOT use tab spaces.

5. If using photographs, either insert them into the paper in the required position, or use a place-marker with caption to indicate the position. Photographs should be supplied in .jpeg format, saved at 300dpi and be at least 1024x768 pixels in size. Clearly identify the images and how they relate to the paper (Photo 1, Figure 2, etc) and where necessary indicate the correct orientation. If you need further help with this, please contact us.

6. Diagrams/line drawings, etc., should be supplied in .bmp, .tiff or .gif format, at a large file size. Please do not create diagrams directly in your paper using the word-processing programme, e.g. Word, as they do not transfer clearly enough to print quality for the publication.

7. Any diagrams and photographs, etc, used in the article should include a brief explanatory caption/legend, indicating clearly to which item it refers, e.g, Photo 1, Figure 2, Table 3.

Proofs

Proofs will be returned to the authors for checking. Please make any corrections and return to the Managing

Editor with the least possible delay.
Authors should pay particular attention
to the checking of numerical matter,
tables, lists of names and references.

Conditions

By submitting a paper to the *Journal of
Practical Ecology and Conservation*
the author implies that it is original
unpublished work, not being considered
for publication elsewhere, and that the
copyright for the article is accepted for
publication. The copyright includes the
exclusive rights to reproduce and
distribute the article. Copies for
educational or academic research
purposes are allowed.

Correspondence concerning editorial
matters should be addressed to:

Professor Ian D. Rotherham

Managing Editor

Wildtrack Publishing

Venture House

103 Arundel Street

Sheffield

S1 2NT

info@hallamec.plus.com

Journal of Practical Ecology and Conservation Vol . 8 (2) 2009

HEC Associates Ltd.

We offer a professional, practical and cost-effective approach to a range of environmental education and conservation consultancy services through our team of associates.

Our full range of *training, education and interpretation services* for voluntary groups etc. includes:
- Running professional workshops, seminars and conferences;
- Organising and running 'in house' and bespoke training workshops;
- Organising public and community environmental events;
- Providing a membership support service;
- Providing a research, design and production service for leaflets, booklets, interpretation boards and display panels;
- Providing a research, design and production service for education packs (Key Stages 1 to 4);
- Providing oral history interviewing and transcription services; and
- Providing training packages for community groups in oral history techniques, website design, landscape surveying and interpretation.

Our *environmental and ecological conservation consultancy services* include:
- Protected species (Great Crested Newts, Badgers, Watervoles, Bats) surveys and advice;
- Invasive and problem species surveys and treatments;
- Japanese Knotweed surveys, confirmation, advice and treatment;
- Environmental Impact Assessment (EIA) scoping reports;
- Environmental Statements;
- Ecological Impact Assessments;
- Landscape characterisation and impact assessments (including light and noise impacts);
- Review of EIAs and other environmental documents;
- Assessment of sites and developments against standard guidelines;
- Provision of non-technical advice to developers and the general public;
- Advice on appropriate responses to proposals;
- Preparation of Local Development Framework documents;
- Provision of an environmental land management, planning and advisory service;
- Preparation of Proofs of Evidence and provision of expert witness(es) at Public Inquiries;
- Baseline habitat and vegetation surveys and monitoring for EIAs and site management including: phase 1 and extended phase one habitat surveys; hedgerow surveys; and phase 2 quantitative surveys; National Vegetation Classification (NVC) surveys; Invasive plant surveys; Aquatic plant surveys – macrophytes and microphytes; Lower plant surveys – mosses, lichens; Fungal surveys; Trees and woodland surveys.
- With our experienced team of fully licensed operatives we carry out, management and eradication of notifiable weeds such as Japanese Knotweed, Giant Hogweed, Creeping Thistle and Ragwort; and management of weed species such as Rhododendron, Himalayan Balsam and Buddleia.
- Produce site management plans covering ecological, archaeological and amenity use of sites; and
- Provide a practical service for habitat creation and maintenance.

Our *Tourism and Leisure consultancy services* include:
- Recreation studies, audits and assessments;
- Economic assessments;
- Socio-economic profiling.

Contact Details
For more information about our consultancy services or to discuss your project requirements, please contact us tel: 0114 272 4227 or send an email to info@hallamec.plus.com